The DAVID SUZUKI Reader

The DAVID SUZUKI Reader

A Lifetime *of* Ideas *from a* Leading Activist *and* Thinker

DAVID SUZUKI

FOREWORD BY BILL MCKIBBEN

GREYSTONE BOOKS

Douglas & McIntyre Publishing Group

Vancouver/Toronto/Berkeley

04 05 06 07 08 5 4 3 2

Greystone Books
A division of Douglas & McIntyre Ltd.
2323 Quebec Street, Suite 201
Vancouver, British Columbia
Canada V5T 4S7
www.greystonebooks.com

David Suzuki Foundation
2211 West 4th Avenue, Suite 219
Vancouver, British Columbia
Canada V6K 4S2

National Library of Canada Cataloguing in Publication Data
Suzuki, David, 1936–
The David Suzuki reader / David Suzuki.

Includes bibliographical references and index.
ISBN 1-55365-022-0

1. Human ecology. 2. Social ecology. 3. Environmental ethics. I. Title.
GF80.S84 2003 304.2 C2003-910906-2

Library of Congress Cataloging-in-Publication Data
Suzuki, David T., 1936–
The David Suzuki Reader/David Suzuki.
p. cm.

Includes bibliographical references and index.
ISBN 1-55365-022-0 (pbk.: alk. paper)

1. Human ecology—Political aspects. 2. Human ecology—Economic aspects.
3. Environmental ethics. 4. Philosophy of nature. 5. Consumption (Economics).
6. Avarice. 7. Globalization. I. Title.

GF50.S89 2004
304.2—dc22 2003049316

Editing by Yvonne Van Ruskenveld and Nancy Flight
Cover and text design by Val Speidel
Cover photograph by Chick Rice
Printed and bound in Canada by Friesens
Distributed in the U.S. by Publishers Group West

Greystone Books is committed to reducing the consumption of old-growth forests in the books it publishes. This book is one step towards that goal. It is printed on acid-free paper that is 100% ancient-forest-free, and it has been processed chlorine-free.

We gratefully acknowledge the financial support of the Canada Council for the Arts, the British Columbia Arts Council, and the Government of Canada through the Book Publishing Industry Development Program (BPIDP) for our publishing activities.

Contents

ECONOMICS AND POLITICS 89

SCIENCE, TECHNOLOGY, AND INFORMATION 155

Foreword

THIS COLLECTION OF ESSAYS AND ARTICLES IS FULL OF INSIGHTS about the world. But read carefully, they are even more full of insights about the author, one of the continent's most remarkable men. And most unlikely men.

Begin with the drama that shaped his early life—his internment along with other Canadians of Japanese descent during the Second World War. Unjust, certainly, but it offered David Suzuki a pair of gifts. The most obvious was an introduction to the splendor of the natural world in the B.C. backcountry. Without a school to "imprison" him, the six-year-old Suzuki roamed the mountains that would in later days become Valhalla Provincial Park. Glory like that leaves its mark—nature in its wildest moods and most majestic expressions has been a touchstone for him since.

But the internment offered another gift, one that didn't develop until much later. He has spent his life open to the idea that power can be abused—any power, the power of a government or the power of a chainsaw or the power of a driftnet. That intuition rescued him from a life as a prominent scientist—as a young biologist, Suzuki won every award there was to win and seemed set for a life at the lab bench. But he was a geneticist, and eventually it dawned on him that even the power of an idea could be abused—dawned on him that it was his genes that that had turned him into a prisoner before grade one.

And so he began his second life, as a communicator, the preeminent explainer of the natural world to television viewers around the planet. Before long, the CBC was peddling *The Nature of Things* in dozens of countries and Suzuki had emerged as a kind of Jacques Cousteau without a wetsuit. He was remarkably good in this role—in large measure, as this collection makes clear, because of his inordinate curiosity about everything that happened on the surface of the globe. Giant fungi, tiny fruit flies. Because he was a scientist, he knew how to make sense out of what he was seeing (and knew enough, too, to reject scientific "certainties" offered as explanation).

He had that unteachable warmth that allowed audiences to identify with him, the only explanation for a television career that is numbered in decades.

But he was rescued from this second career too—rescued from being nothing more than a talking head pointing at the cheetah as it consumes its kill. He emerged as Canada's nature authority at the precise moment that nature itself passed into its crisis, and he possessed too much sanity and too much backbone to ever pretend it wasn't happening. You might say he became radicalized, but in a deeper sense it was the world that turned radical, and he was simply honest enough to point it out—the clear-cut forests of Sarawak, the strip-mined deep seabed, the atmosphere filling with carbon dioxide till scientists now predict that the Earth will be 3 degrees warmer by century's end. Easier by far not to notice this, of course, or at least not to pay it more than passing attention—doubtless the CBC would have preferred someone a little less vigorous.

Even here, however, Suzuki pushed on. Most environmentalists content themselves with pointing out the practical folly of our treatment of the Earth—that by tearing down the rain forest we will obliterate species that might someday treat our cancers, that by clogging the atmosphere with greenhouse gases we will cost ourselves large sums of money. All true, but also the easy way out.

Suzuki has committed himself to asking a deeper set of questions, reflected time and again in these essays. He questions the "rationalism" of the academy, even of his beloved University of British Columbia. He questions the very medium—television—that won him his audience. And, seditiously, he questions the global economic system that underwrites it all. Endless growth is impossible, he declares—a simple formulation that defies the basic current of our time. Global economics is "perverted." He makes these points before a meeting of the World Bank and is greeted with a standing ovation but also a pointed question. "Why don't you work with us?" one bureaucrat asks. "It's just a matter of getting the pricing right." Wrong, Suzuki insists. Important as markets are in offering a solution to our various woes, "the most important things are beyond price. How much is your mother worth? Or your sister or your child?"

Other factors have helped produce this unlikely man. Living on the fringe of the American empire was an obvious piece of luck, allowing him to see close up but clear-eyed the ways the world was being reshaped. Encountering the First Nations people of Canada at the moment that they began their renaissance helped him see nature with new eyes. Whatever the tributaries, though, they have merged into a mighty river of a man, as this book makes clear in a way that perhaps his more focused efforts cannot.

"People often ask me, What is the most urgent environmental issue confronting us?" he writes. "I believe the overarching crisis resides in the modern, urban human mind, in the values and beliefs that are driving much of our destructiveness." That is not sentimental bosh—that's the considered opinion of a man who stands on firmer ground than most of us. In the essays that follow, diagnosis overlaps with prescription, description with passion, all of it filtered through a love so strong that it will not let us off easily. May they help make more of us into unlikely citizens of a needy planet.

Bill McKibben

Preface

AS A SCIENTIST, I WAS NOT TAUGHT TO FEEL A RESPONSIBILITY TO discuss moral, ethical, or social ramifications of new discoveries. In fact, my fellow scientists regarded the popularization of science as a rather vulgar activity undertaken by those who couldn't cut it in research.

I became interested in the interface between science and society when, as a university professor, I discovered how genetics and the seminal experience of my life, during World War II, had intersected—my family had been treated as a threat to Canada because of our ethnic background. As a consequence, I entered a new (for me) area of journalism in radio and television. In this capacity, I was able to explore a far wider range of topics and ideas than I would have paid attention to as a geneticist. So it seemed a natural evolution to begin making observations and expressing opinions in print in articles written for magazines, newspapers, and books.

Many of the pieces I wrote were published in books that became bestsellers (*Inventing the Future,* 1989; *Time to Change,* 1994; and *Earth Time,* 1998). Looking back at these essays, written over a period of twenty-five years, I am surprised that many remain relevant today. So in the best environmental tradition, I am recycling the best of these along with new essays never published in a book before in the hope of finding a new audience for the ideas.

I'm very grateful to Greystone Books for pressing me to gather some of the essays I've written for a kind of "Best of..." book. Yvonne Van Ruskenveld and Nancy Flight performed the heroic task of actually reading a large number that I sent to them, making the first cut, and ordering them. I am deeply grateful to them for that.

Over the years, I could see our world being shattered as reports were reduced to snippets of information. Almost every time I listened to the news or read a paper, I felt a need to make sense of stories by fleshing them out or setting them in a broader context. I became obsessed with writing articles, and the person who carried the brunt of that preoccupation was my wife, Dr. Tara Cullis. Thanks, sweetheart, I owe you big time.

INTERCONNECTIONS

THROUGHOUT HISTORY, PEOPLE HAVE ALWAYS UNDERSTOOD THAT we are deeply embedded in and utterly dependent on the natural world. In stories, songs, and dances, cultures around the globe have celebrated being part of their surroundings. In a world where everything is connected to everything else, any action has repercussions and so responsibilities accompany every deliberate act. Acknowledgment of that responsibility has also been explicit in the rituals of every society.

That sense of being an intimate part of nature has been shattered over the past few centuries. Science brought forth a different way of looking at the world. If the cosmos is an immense mechanical construct, then by focusing on the parts of this machine—the cogs, wheels, and springs—one might be able to understand how the machine works; like a giant jigsaw puzzle, all the parts could be fitted together to explain the whole. Reductionism, the focusing on parts of nature to explain the whole, was repudiated by modern physics but continues to be the underlying assumption in most of biology and medicine. Reductionism has fragmented the way we see everything and obliterated the rhythms, patterns, and cycles within which the parts operate.

As human numbers have exploded, more and more of the world's people have spent their entire lives in a period of unprecedented growth and change. To them, this is the normal condition of humanity and it is expected—indeed, it is demanded—by those who know no other way.

In the twentieth century, there was a remarkable transition from villages and rural communities to big cities. In an urban environment, it has become easy to assume that human technology, economics, and industries create our habitat and fulfill our material needs, thereby enabling us to escape the bounds of nature.

At the same time, information has mushroomed, assaulting us from all sides in print and electronic media. But the bulk of the information that hits us is junk—advertisements, pornography, mindless entertainment—and does nothing to inform us about the important matters in our lives. Even the so-called news reports provide only snippets of information, devoid of explanation, history, or context that might explain why they even matter.

A global economy continues the process of disconnection by treating nature and all its services as something outside economics. Economists assume that the atmosphere, water, soil, and biodiversity that are crucial to all life are "externalities" to economics. For example, the natural services rendered by a tree, including exchange of carbon dioxide for oxygen in the air, prevention of soil erosion, influence on weather and climate by transpiration, and provision of habitat for other species, have no economic worth, whereas cutting the tree down for lumber or pulp does. So human intrusion and exploitation contribute to the global economy, whereas nature's activities in keeping the planet habitable are "worthless."

Taken all together, these factors have severed the sense of interconnectedness and with it, any feeling of responsibility for what we do. The challenge of the twenty-first century is to recognize that our biological nature determines our most basic needs, which all come from nature.

Catching an Epiphany

I WAS BORN IN VANCOUVER, BRITISH COLUMBIA, IN 1936. MY EARliest childhood memory is one of almost unbearable excitement—my father and I went to a store to buy a tent so that we could go camping. One of the salesmen set up a pup tent right there in the store, and Dad and I scrunched in and lay down together. Dad wrapped his arms around me and held me close as I squirmed in uncontrollable anticipation of our first camping trip.

My recollection of those early years in British Columbia is mostly snippets of memories around camping and fishing trips. On my very first fishing trip, at age four, we hiked in to Loon Lake, more a large pond than a lake, and while Dad fly-fished, I sat on the dock with a small rod and reel and a tin of manure worms, pulling in one small trout after another until I had my limit. I remember sitting in a rowboat as Dad trolled for sea-run cutthroats around Stanley Park; we also jigged for halibut off Spanish Banks, cast for sturgeon near the mouth of the Fraser River, and hiked up the Vedder River to catch steelhead and Dolly Varden.

Dad loved to take children fishing and turn them into avid anglers and outdoors people. His philosophy was to make sure a first-time fisher caught fish. Size didn't matter; it was the catching that hooked kids. To this day, adult strangers approach me to tell me how much they appreciated those fishing trips with Dad and how they are doing the same with their children and their friends. Throughout my childhood and adolescence, fishing continued to provide some of my most memorable experiences.

After the Japanese attack on Pearl Harbor, twenty thousand Japanese Canadians, most Canadian citizens by birth or naturalization, were rounded up under the iniquitous War Measures Act, which legitimizes the suspension of all rights of citizenship at times of ill-defined threat. So my Canadian-born parents and siblings and I were rounded up and shipped out of Vancouver with seventy pounds of luggage each. Dad was sent to a "road camp," where he worked on the Trans-Canada Highway, while my mother, my two sisters, and I were evacuated to the interior of British Columbia.

As a boy of six, I was shielded from the upheaval and my parents' distress, and everything seemed like a grand adventure. We went on a long train ride to be interned in an old ghost town that was a relic from a silver rush in the 1890s. I didn't go to school for a year because there were no facilities or teachers. We were squeezed into a tiny room in a decaying hotel crawling with bedbugs, where we shared toilets, cooking facilities, and baths that were so big I learned to swim in one. But none of this mattered to me because the internment camp that was to be home for three years was located in a magical place. We were at the end of a long, narrow lake—Slocan Lake—at the bottom of a valley whose western slopes would become Valhalla Provincial Park three decades later.

Without school to imprison me, I roamed the shores of the lake and rivers and the forests on the surrounding mountains like a young voyageur. The lake was filled with rainbow trout, whitefish, kokanee (landlocked sockeye salmon), squawfish, suckers, and chubs, and I fished for any of them indiscriminately.

By watching older boys and by trial and error, I learned to use caddisfly larvae and helgramites as bait; I found that in early morning when the dew was still heavy, grasshoppers were stiff and torpid and a cinch to catch. A year later when Dad joined us, we explored the mountain creeks to fish for the abundant trout. In the camps, Japanese internees were forbidden to fish, but to us fish were a staple like rice and we carried on a cat-and-mouse relationship with the authorities as we fished at every opportunity while trying to avoid being caught. In those mountains, we encountered wolves, black bears, coyotes, and porcupines; carefully avoided huge clumps of spiny devil's club; and gathered bags of prized pine mushrooms (*matsutake*) in the fall. This was biology as it should be learned, firsthand in the wild, joyously and effortlessly.

As the end of the war approached, those Japanese-Canadian families like mine who chose to remain in Canada were expelled from British Columbia, and we ended up in Leamington, Canada's southernmost town, on Lake Erie. This area of southern Ontario was intensive farming country, with flat fields and ditches running along country roads and small wood lots on private property, a radically different environment from British Columbia. But that very difference provided new opportunities for exploration. Spurred by Dad's excitement about discovering entirely new species of fish and armed with dip nets, my sisters and I investigated every ditch, creek, and pond within biking range. We kept bottles of sunfish, small catfish, baby turtles, and minnows. We fished for channel catfish, sheephead (drum), silver bass, and smelt, each savored for its distinctive taste and texture. And we discovered Point Pelee, a jewel perched on the southernmost tip of Canada. There the marshes were filled with birds, reptiles, insects, and fish, and the lakeshore was littered with fossils and dried carcasses of birds and fish. Point Pelee would later become a national park that is one of the most popular gathering places for birds and their watchers.

In 1949, we moved to London, Ontario, a city of seventy thousand people that was growing explosively and would increase fivefold in the next half a century. Yet even here there was much to fascinate me. My family had been wiped out financially by the evacuation, incarceration, and expulsion, so we gathered edible plants and caught fish to supplement our diet. Even though the Thames River ran right through the middle of London, it was a fabulous place to fish. I came to know every pool and riffle along a 2-kilometer (1-mile) stretch near our house. There were plenty of black bass, carp, catfish, and suckers, and at certain times of the year, the river was jammed with spawning pike, silver bass, and pickerel. I learned where and how to catch frogs, crayfish, minnows, and leeches to use as bait.

Only a kilometer (½ mile) from our house was an immense swamp that held endless attractions. I was a loner. The war years had left me with an overriding sense of inferiority, and I anticipated rejection because I was Japanese. To make matters worse, my feeling of alienation was strongest just as I was in the throes of puberty. This was during an era when virginity was still prized, so it was impossible to do much about my sexual fantasies. That swamp was my salvation—all my hang-ups, fears, and frustrations fell away whenever I biked to that marsh. I was an avid insect collector, and the water was filled with bizarre creatures—whirligigs, water striders, and boatmen. It was in that enchanted place that I first spotted a bittern with its beak pointing straight up, trying to blend in with the surrounding reeds. And each spring, the swamp reverberated with the sexual calls of frogs that proved irresistible to mates and me. I'd return home triumphantly carrying jars of salamander and frog eggs that I could then watch metamorphose into tadpoles.

My father's enthusiastic embrace of new fishing places and species had shown me there were worlds to be experienced wherever one lived, and I found that to be true when I finished high school and moved from Ontario to Massachusetts, Illinois, Tennessee, and then Alberta, where I became an assistant professor of genetics at the University of Alberta. Each place offered new environments and organisms to explore. But unlike the wilderness of British Columbia, where I had bonded with nature as a young boy, all of my postwar experiences were in areas where people dominated the countryside, and I had to search for the pockets of nature that still flourished.

After an especially cold winter in Edmonton, I accepted a position at the University of British Columbia, in Vancouver, in 1963, returning at last to the city of my birth. With a young family, I repeated the example of my parents, planning camping and fishing trips for my children so that they could experience nature. Every weekend, we would try to get out into a new area, following leads provided by others, driving along country and logging roads around Vancouver to reach isolated rivers and lakes.

I had been at UBC for about a year when I heard about a logging road near Squamish that would take us to a river that was supposed to contain good-sized rainbows. So one Saturday I loaded my children, Tamiko and Troy, into the car and took off for a day trip. Passing Squamish, we left the pavement for a dirt road and soon encountered a sign announcing that we were entering forest company land, where logging trucks had the right-of-way. The road was in excellent condition and wound through the hills for kilometers. In the area being actively logged, the forest had been cleared and the debris that was left gathered into great piles to be burned. Having spent most of my life in the east, I wasn't especially disturbed by the logging; after all, I loved working with wood in carpentry and used a lot of paper at work, and I knew that forestry was the engine of the B.C. economy. Besides, the logging roads enabled me to get into remote parts of the province.

When the road finally neared the river, I drove up a hill and found a level spot on the shoulder. We put on our day packs, picked up our fishing gear, and set off. All around us was a combat zone where the soil had been churned up by the tracks of heavy machines, and all that remained of the immense trees were huge stumps and roots that projected at garish angles among the slash. From the top of the hill, the logged-out clearing had looked deceptively smooth and easy to traverse, but once we had left the roadside and started to descend the hill, it became tough slogging to get by the debris. Time after time, I was forced to hoist the children up over obstructions. What I had thought would be an easy ten-minute hike turned into an hour, but once committed, I wasn't about to give up. I kept bantering with the children and playing games with them as we worked our way across the clearing. I was far too focused on the challenge of traversing the clearing to reflect on the fact that this *was* a war zone where human economic demands were conflicting with the continuation of the community of life making up this ecosystem.

It was a sunny day and I soon found myself sweating profusely, kicking myself for not bringing any water, and worrying about the children. After much puffing and unjamming of the rods I was carrying from branches and debris, we finally reached the trees at the edge of the logged area. Stepping out of the glare and heat of the clearing and into the dark, cool cathedral of trees was an absolute shock, like stepping from a hot city street into an air-conditioned building. Embraced by the cool shade of the trees, we inhaled the damp, musky odor of vegetation and decaying tree carcasses. We were enfolded in silence. The children immediately stopped bickering and complaining and began to whisper just as if they were in a church. As our eyes adjusted to the shade, we saw that the forest floor was cloaked with moss that smoothed everything into an undulating carpet. The bodies of great leviathans of fallen trees cold be seen in outline under the moss, in death nurturing a community of huckleberry, sword ferns, and small trees.

As we searched for a trickle of water to drink, the crackling of branches under our feet was muffled by the vegetation. High above us, the canopy stretched to the sky with green branches and needles jockeying for a place in the sun and allowing an ever-shifting filigree of speckled light onto the forest floor. As terrestrial creatures, we could only wonder at the drama of life cycles and predation taking place in the nooks and crannies of the branch tips, needles, and leaves of the canopy and in the soil community hidden beneath our feet. Tamiko, Troy, and I joined hands and reached around the circumference of one of the trees, not even reaching halfway. Those giants must have been hundreds of years old.

I was dumbstruck. Nature had always been my touchstone, but I had spent much of my life in Ontario, where forests had been heavily affected and altered by people. There trees had been extracted, creeks rerouted, and the soil cultivated or developed. This was a forest shaped by the forces of nature for ten thousand years, a community of life where death gave birth to new life in an endless recycling of nutrients through the countless species that make up a forest. We had stepped into it from the edge of industrial logging, which would soon transform it into something infinitely simplified and unrecognizable. In those few minutes that my children and I had entered into the forest temple, I had recognized the terrible hubris of the human economy. To transform this matrix of life-forms, soil, water, and air into a war zone where soil, air, water, and life were so degraded was a travesty of stewardship and responsibility to future generations. I didn't articulate it that way at the time. I only knew in a profoundly visceral way that industrial logging was not right, that the magnificent forest we had entered was an entity far beyond our comprehension and was worthy of respect and awe.

I had been set up to have that inspirational encounter with an old-growth forest after reading Rachel Carson's seminal book, *Silent Spring*, in 1962. Years later, I would encounter First Nations people who would educate me about kinship with other species and the way in which we are all interconnected and interdependent. But that encounter with an ancient forest on the edge of a clear-cut was my moment of enlightenment.

Today, when my grandchildren beg me to go fishing with them, I can't take them to Spanish Banks or the mouth of the Fraser or the Vedder River or other places where my father took me. I can't go back to fish in the Thames River, which is so polluted that people recoil at the thought of eating anything caught there. I can't return to the swamp that soothed me during my adolescence, for it is now covered with an immense shopping mall and parking lot. And the forest that was my epiphany was felled within weeks of my visit there. What remains is my conviction that we must rediscover our biological place and learn to live in balance with the natural world that sustains us.

What Can I Do?

PEOPLE OFTEN ASK ME, "WHAT IS THE MOST URGENT ENVIRONMENTAL issue confronting us? Is it climate change, species extinction, toxic pollution, or deforestation?" The only honest answer is, "All of the above and more are serious ecological issues, but no one knows which one might trigger an irreversible and catastrophic collapse in the planet's life support systems." There is no single act that will somehow avoid the looming ecocrisis.

I believe the overarching crisis resides in the modern, urban human mind, in the values and beliefs that are driving much of our destructiveness. Throughout the history of our species, human beings have understood that we are a part of nature, in which everything is connected to everything else and nothing exists in isolation. Rachel Carson pointed out that insecticides sprayed to kill pests inadvertently end up affecting fish, birds, and human beings. So every deliberate act carries the responsibility to think beyond the immediate issue and consider the whole system.

Today it's difficult to recognize our continuing connection with and dependence on nature. So try this thought exercise: Imagine that scientists have created a time machine and we travel back four billion years, before life had evolved on the planet. Instead of unlimited resources and opportunity, we discover a place inhospitable to human life. The atmosphere is poisonous, rich in carbon dioxide and devoid of oxygen. Because there is no life, water is not filtered of toxic materials by plant roots, soil fungi, or microorganisms. There is nothing to eat in the prelife world because every bit of our food is composed of plants, animals, and microorganisms. Even if we have brought a stash of food and seeds in the time capsule so that we can stay a while, there is nowhere to grow food because soil is created by the mixture of molecules from life-forms with the matrix of sand, silt, and clay. There is nothing to burn to create heat because all fuels—coal, gas, oil, wood, and peat—are created from life. Even if we have brought paper and wood to make a fire, the absence of oxygen precludes flame anywhere on Earth. So the four sacred elements—earth, air, fire, and water—that traditional people tell us sustain all life, are created and cleansed or replenished by the web of life that we tend to call nature.

The way we see the world shapes the way we treat it. If a mountain is a deity, not a pile of ore; if a river is one of the veins of the land, not potential irrigation water; if a forest is a sacred grove, not timber; if other species are our biological kin, not resources; or if the planet is our mother, not an opportunity—then we will treat each one with greater respect. That is the challenge, to look at the world from a different perspective.

Arrival of an Alien

THE SURVIVAL STRATEGY OF OUR SPECIES IS THE POSSESSION OF A large and complex brain that has endowed us with the ability to project ahead into the future and anticipate the possible consequences of our actions. As a result of this foresight, our species is unique in being able to make choices. We can recognize options, weigh their hazards and benefits, and then deliberately choose the action that will maximize the likelihood of benefit while minimizing dangers. This evolutionary strategy has succeeded wonderfully for *Homo sapiens* and has given us a position of unparalleled dominance on the planet. So it is ironic that in a time of enormous amplification of human brainpower through computers, telecommunication, scientists, and engineers, we no longer seem able to do what our ancestors did routinely, that is, to assess the hazards that confront us and to choose the best option for long-term survival.

As biological creatures with organs to inform us about the state of our surroundings, we are very good at reacting when an immediate crisis hits. A car accident, fire, earthquake, flood, or storm leaves us no choice but to do the best we can to survive and recover. But it's far better to anticipate possible difficulties or hazards and act to minimize the likelihood of their occurrence.

Imagine that an immense alien from another galaxy arrives on this planet. Each of its two feet is about an acre in area, and it runs across the planet taking one step every second and crushing everything under each footstep. The creature has an insatiable appetite, draining lakes and rivers to slake its thirst, capturing huge quantities of fish in the oceans, ripping open mountains to get at minerals, spewing vast clouds of toxic fumes into the atmosphere from its mouth, and fouling the land and water with poisons excreted from its other end. Faced with such a monster, the entire world would be united in declaring a global emergency and launching a massive response that would dwarf the current mobilization to fight terrorism. Everything would be done to stop the creature from its destructive rampage.

The collective impact of human activity is comparable to that alien's, and it is a puzzle why we aren't responding appropriately. Instead, we are stopped by objections that the destruction is not really that bad, or the costs of stopping the rampage are unacceptable, so we let the monstrous activity continue. We are no longer using our foresight to wend our way safely into the future.

I believe that one of the reasons we seem unable to respond to the threats of human activity is that we no longer see ourselves as part of the natural world. As a species, we have been able to survive in surroundings as varied as high alpine regions, the Arctic, deserts, grasslands, forests, and wetlands. Our survival in such different circumstances has required a profound working knowledge of our environment. But more and more people dwell in the human-created environment of cities, where it is easy to accept the illusion that we no longer need nature because trade enables us to exist unrestricted by the limited productivity of our immediate surroundings and a strong economy enables us to have all of the trappings of modern life, including a clean environment.

When a species like ours becomes so numerous and demanding through technology, consumption, and economics, the repercussions for the rest of life on Earth are immense. Our transformation into a major force affecting the biological, physical, and chemical features of the planet has happened so suddenly that most of us have not yet recognized that for the first time in history, we must be aware of and concerned about the collective impact of our entire species.

The challenge of our time is to see that humanity as a whole has become that rampaging alien, to use our foresight and judgment to undertake preventive measures, and to marshal the kind of response we would get if there really was an invasion from outer space.

London in My Life

WITH RELUCTANCE AND ASTONISHMENT, I HAVE BECOME A MEMBER of elders in society. As I reflect on the changes that have occurred during the brief span of my family's and my own life on Earth, it becomes clear that the enormous changes made during that time cannot be sustained. All across the planet, people in towns and cities undergoing explosive growth in population and economic development have reason to pay attention to the experience of their elders.

My grandparents were driven out of Japan by poverty at the beginning of this century and came to Canada to seek their fortune. They had no intention of staying in what they considered a primitive and backward country. All they wanted was some of its wealth to take back home. My grandparents were aliens in an unfamiliar landscape with which they had no historical or cultural link, let alone a sense of reverence for its sanctity. To them, Canada represented an opportunity; the land was a commodity full of resources to exploit. My grandparents became a part of a massive assault on the "New World" initiated by Columbus's arrival and causing vast ecological and human catastrophe.

Following World War II, my family moved to Ontario, the industrial heartland of Canada and the most populous province in the country. First in Leamington, then in London, I grew up in a land named after the homelands of the European settlers. There were few reminders that this area had long been occupied by people with proud histories, people who had been mistakenly labeled "Indians" by the newcomers. But lumped together as red Indians were dozens of nations, including Algonquins, Mohawks, Cree, and Ojibway. Today, most aboriginal people of Canada are invisible, sequestered on reserves or hidden through forced assimilation.

In London, we lived on the northwest edge of town next to the railway tracks, along whose banks I would pick asparagus in the spring and hunt for insects in summer. One year I worked on a vegetable farm only a kilometer (half-mile) or so down the railway line. A few blocks east of our house was the Thames River, where I discovered an incredible treasure—a softshell turtle!

Bicycling west on Oxford Street, I would quickly run out of pavement and hit the gravel road. In about twenty minutes, I'd be at my grandparents' 4-hectare (10-acre) farm at the end of Proudfoot Lane. But first I'd always stop at the large swamp beside the road to look for frogs, snakes, and damselflies. Many times I returned home with boots full of mud and bottles containing frog eggs and dragonfly larvae. The woods surrounding the swamp always beckoned with the promise of a glimpse of a fox, skunk, raccoon, or owl.

My grandparents' farm was a child's paradise. Besides large vegetable and berry patches to be raided, there were several hundred chickens to be fed, eggs to be gathered, and fences to be mended. At the end of the fields, a creek ran year-round. That was where I dipped for darters, discovered freshwater clams, and hunted snails. In the fields, pheasants tooted like trains, groundhogs sunbathed in front of their burrows, and hawks skimmed above the ground in search of rodents.

In the thirty-five years since my boyhood, the Thames River has been saturated with industrial effluent and agricultural runoff accumulating along its length. The river was too convenient for dumping garbage and chemical wastes. Now there are few clams, crayfish, or minnows to be seen. Londoners today recoil at any suggestion of eating fish from the Thames or asparagus from the tracks.

When I arrived in London in 1950, its population was just over 70,000. Five years later, we were proud when the city passed 100,000. By 1960, it had almost doubled to over 185,000 and reached a quarter of a million ten years after that. Today, London boasts 350,000 people. This spectacular rate of growth was accompanied by a booming economy and a sense of civic pride. But at what cost?

The road to my grandparents' farm is now a wide highway, with the city extending all the way to the village of Byron. My grandparents' farm is occupied by a cluster of high-rise apartments, and the creek has been tamed to run through culverts. My beloved swamp is covered by an immense shopping mall and parking lot, and the woods beside it have given way to a huge housing complex. Along the Thames River and all around the city, once productive agricultural land has been converted to housing subdivisions.

Within my lifetime, the ecological devastation has been massive. But when my grandparents immigrated to North America, the real holocaust had already occurred. Only two hundred years ago, Ontario was covered by a dense, ancient forest, the plains of the Midwest reverberated under the hooves of 60 million bison, and the skies were darkened for days on end when billions of passenger pigeons passed by. By the beginning of the twentieth century, they were all gone, yet we have learned little from that unprecedented ecological annihilation and continue our destructive rampage; we can see the destruction going on before our eyes.

In the topsy-turvy world of economics, farmland, swamps, woods, rivers, and ponds adjacent to expanding cities acquire value that makes it inevitable for them to be developed. So the animals and plants that belong there disappear, leaving our children to grow up in an increasingly sterile human-created environment. And with diminished opportunities to experience nature, our future generations become all the more estranged from the real systems that support their lives.

My hometown of London is a microcosm of what has been happening around the planet, but especially in the New World and especially since World War II. Seen from a plane above Canada today, the country is crisscrossed by geometric straight lines of highways and rectangles of clear-cuts and agricultural fields. Everywhere the imprint of human beings has been stamped on the land with a mathematical precision that pays no attention to geographic and biological realities. We act as if our political subdivisions of the land are meaningful and fail to observe the realities of "bioregions," ecosystems and watersheds to which living things have evolved and fit.

Our alienation from the land is so great that we have no sense that it is sacred or that our ability to exploit it is a great privilege accompanied by responsibility. Impelled by our faith in our technological prowess and scientific knowledge, we assault the planet as if it is limitless and endlessly self-renewing. Like an exotic species introduced to a new environment, we feel no natural restraints, only the deadly belief that all of nature is there for us to use as a resource in any way we wish.

This story has been repeated in different parts of the world. Driven by a profound disconnection from the land, newcomers have sought to tame it and its human and nonhuman occupants. The combination of technological power and the Western attitude of rightful dominion over nature has been unstoppable. That has been the legacy passed on to the present time.

Seen from another perspective, beginning with respect for the unique flora and fauna of the continent and extending to the indigenous people whose cultures were so exquisitely evolved to live in rich harmony with the land, the technological optimism and economic greed of the invaders become a policy that is shortsighted and arrogant.

It can be argued that one of the great tragedies that led to the current crisis in wilderness destruction was the attempt by colonizing peoples to re-create their familiar European surroundings in alien lands. In Canada and Australia, forests, grassy plains, and swamps were forced to resemble bits of home. And the introductions of species such as sparrows, foxes, and rabbits were ecological catastrophes.

Each visit to my childhood roots becomes a bittersweet mix of memories that remind me of the price we have paid for the way we now live.

Galápagos

HOMO SAPIENS IS A TRULY GLOBAL ANIMAL. OUR ADAPTABILITY, made possible by the inventiveness of our brain, has enabled us to occupy every continent. There is nowhere on Earth that we haven't been. And we have invaded that hallowed biological laboratory that shaped Darwin's thinking. Once remote hiding places for pirates, the Galápagos Islands are magical jewels attracting hordes of ecotourists.

The name Galápagos conjures up other words—HMS *Beagle,* Charles Darwin, finches, evolution. To a biologist, a visit to these fabled islands is a pilgrimage to the source of the inspiration for the great unifying concept in the life sciences. The remote equatorial archipelago, long protected from human habitation by its isolation, was an evolutionary laboratory seemingly made just for the observant scientist. Today, the islands are an Ecuadorian national park that allows the privileged visitor to take a trip back through time.

On arrival, my first impressions were of the animals that are present in awesome profusion—the iguanas, frigate birds, boobies, sea lions, flamingos, and tortoises. Birds, reptiles, and marine mammals, often in astonishing numbers, share overlapping territory, with remarkably little overt aggression. The most emotional part of the Galápagos experience for me was the animals' complete lack of fear of humans. It was profoundly humbling to be ignored as a nonthreatening part of the surroundings. Based on what we have done to creatures elsewhere on Earth, Galápagos animals should flee from us in terror. I am deeply grateful that they don't.

My second reaction was that the planet has grown so small that we can't escape the evidence and impact of our species. It's not just the bits of plastic and other human-created debris to be seen on every beach; ecotourism itself is the main force that shapes the fate of the flora and fauna here. Two airports allow jets to bring in a torrent of tourists, who support the islands' human communities as well as the Ecuadorian government. As I watched the oil slick from bilge water being pumped from our small boat, I couldn't help thinking that even the most enlightened tourists have an effect. And although ecosystems are resilient, there are limits.

When the second airport was built a few years ago, the ceiling on tourists was raised from 25,000 to 40,000 a year. The village of Puerto Ayora on Santa Cruz Island is exploding with unlimited immigration of Ecuadorians, and the impact of the 5,500 settlers is apparent everywhere. Continued growth of the island population will inexorably put greater pressure on the island ecosystems, and human and nonhuman needs will inevitably conflict.

The basic problem is that our ignorance is simply too vast to allow "management" of complex communities of organisms. The best approach is to be very conservative and tightly control the most destructive element in the islands, namely, us. History suggests that's not likely.

Over tens of millennia of isolation, each island of the Galápagos was an evolutionary opportunity. New species arrived as part of the flotsam and jetsam that blow and wash onto any ocean islands. Most have disappeared; only a few survived. But like North and South America and Australia over the past five centuries, the Galápagos Islands in recent time have been radically altered both deliberately and accidentally by a succession of pirates, whalers, and settlers.

The famous giant tortoises, a source of fresh meat for ocean voyagers, were carted off by the tens of thousands, extinguishing them from some islands. Introduced plants such as elephant grass, guyaba, and wild cucumber have altered the species mix on some islands. Insects such as wasps and fire ants have taken hold and have become major pests for people, while the black-billed ani, a bird introduced to eat ticks and insect parasites on cattle, has become a major competitor with the endemic birds.

But the real disasters have been the mammals such as cats, burros, goats, pigs, rats, and dogs. The government has instituted "control" programs to reduce or eradicate goats and cats by poison, traps, and hunting, but the logistical problems are enormous and each has its associated negative side.

The Charles Darwin Station on Santa Cruz Island supports research and has a breeding program to increase numbers of threatened animals such as tortoises for release back into the wild. But again, human perceptions and priorities based on limited knowledge are being imposed on the islands.

So although a visit to the Galápagos is a sublime experience, it does not provide an escape from the reality of the global ecocrisis. Yet sharing space that is home for other species is spiritually uplifting, filling us with awe, reverence, and indeed, love. These emotional connections could be the beginning of a new attitude that might eventually change the way we live on Earth.

Human Borders and Nature

WE HAVE A REMARKABLE ABILITY TO DEFINE THE WORLD ACCORDing to human needs and perceptions. Thus, although we draw borders to demarcate countries, provinces, or counties, these lines exist only on maps that humans print. There are other boundaries of far greater significance that we have to learn to recognize.

Through the powerful lenses of our *speciesism,* the belief in our superiority over all other forms of life, we take transitory human boundaries far too seriously. Natural barriers and perimeters of mountains and hills, rivers and shores, valleys and watersheds regulate the makeup and distribution of all other organisms on the planet.

You get a sense that there are other ways of delineating territory when you encounter an animal trail in thick underbrush in a heavily wooded area. I've gratefully discovered these narrow paths created over time by large mammals and followed them out of the bush. Often they meander along natural contours like a stream bank, the base of a hill, or rock outcroppings. But they may also branch off unexpectedly, end abruptly, or wander aimlessly according to the mysterious needs of animal priorities. Like animal trails in a forest, the grassy tunnels of rodents in a meadow also wind tortuously for unfathomable reasons. The movement and distribution of animals and plants do not conform to human imperatives.

Aboriginal people speak of "listening" to other biological inhabitants of their territory. They "feel" what the land wants and are respectful of it. We, in urban industrialized societies, have disconnected ourselves from these physical and biological constraints. For speed and efficiency, we remake the countryside around us. We flatten hills, drain potholes and swamps, fill in shorelines, straddle lakes and rivers, and straighten ditches. We impose our will on the land and force it into a form that we want.

It's the same with our political borders. They tend to divide up the country in straight lines without regard for the contours of the land. Local politicians seldom deal with a watershed in its entirety. Instead, garbage dumps, factories, and toxic-waste holding ponds are often put in the county with the most lax regulations, even though the water is shared between neighboring counties.

Our human-created boundaries have become so real that we think that air, water, land, and different organisms can be administered within the limits of our designated jurisdictions. But nature conforms to other rules. Pacific salmon may migrate over thousands of kilometers. Driven by their own impulses, they become prey for people in many countries, yet catch limits are set locally without regard to the total ecological needs of each species. Canadians can't "manage" salmon that leave Canada's shores to feed along other coasts.

Living organisms distribute themselves according to their biological needs. There are no such things as "Canadian" fish or "British Columbia" plants, any more than air, clouds, or oceans can be encompassed by human borders.

Nothing illustrates the insignificance of human borders better than shorebirds. As a biologist, my great passions in life have been insects and fish, but working on a program called "Connecting Flights" for *The Nature of Things,* I got to watch birds.

At one point, we drove south of Vancouver through the most densely populated part of British Columbia, past urban sprawl, shopping malls, golf courses, and farmers' fields. And there in Boundary Bay, an area of ocean mud flats formed at the mouth of the Fraser River, was an incredible biological spectacle—tens of thousands of migratory shorebirds within sight of the airport! Those birds were drawn to that specific feeding ground at just that time of year by an inbuilt genetic clock set through thousands of years of natural selection and evolution. They make a mockery of our political borders.

Every spring and fall hundreds of thousands of shorebirds pass by, and most Vancouverites scarcely notice them. In a spine-tingling display, they skim the water in huge clouds, then explode toward the skies in perfect unison in a kind of celestial cartwheel. The sixty-seven species that nest in Canada include sandpipers, plovers, and curlews. But it's absurd to call them Canadian birds when they travel all the way from the Arctic tundra to South American wetlands and back in a year. Most species cover at least 12,000 kilometers (7,500 miles), and some even go more than 25,000 (15,500 miles)! For some shorebirds, the flight from eastern North America to northern South America is made in forty to sixty hours *nonstop!*

This amazing globe-trotting story begins when dozens of species of shorebirds arrive for the brief Arctic summer. They have to recover their energy and expend more in courtship, egg production, and chick rearing. What makes this incredible journey worthwhile is biological opportunity—widespread nesting sites with little competition and an explosion of food, especially insects. The birds can therefore load up with nutrition and fat, lay eggs, and get their young started, then take off for a destination on the other side of the planet.

During their long stretches of flying, they burn up their fat reserves. Eventually they have to find food or die. Strategically placed along their routes are critical sites where the birds can find abundant energy-rich food. For example, along the rich mud deposits of the Bay of Fundy, shorebirds find and consume an amphipod called *Corophium* at the staggering rate of forty a minute per bird for hours each day. On the way north in May, shorebirds arrive at Delaware Bay in New Jersey at precisely the same time that horseshoe crabs come ashore to lay their eggs. Each female lays fifty thousand eggs, contributing to a rich soup of protoplasm for the birds to dine on. These feeding grounds are irreplaceable. Because the birds lack biological flexibility, they are unable to find new places or to switch sources of nutrition if their feeding grounds are developed, drained, or poisoned.

As the different species of shorebirds fly south, they fan out across wintering grounds that extend from Central America all the way to Tierra del Fuego. In Surinam, for example, hundreds of thousands of shorebirds overwinter on mud flats that have their origins in the Andes and flow north from the Amazon River. Not surprisingly, shorebirds run a gauntlet of human impacts—sport and food hunters, farm pesticides, industrial toxins, urban sprawl and development—that are constantly squeezing the critical sites. They depend on wetlands such as streams, marshes, lagoons, and estuaries that have been radically changed by human activity at the rate of 2 hectares (5 acres) every five minutes ever since the arrival of Europeans in this hemisphere. Only 50 percent of North America's original wetlands remain, and the decline and even extinction of shorebird species have been alarming.

In 1985, to counteract this trend, the Western Hemisphere Shorebird Reserve Network (WHSRN), a coalition of six countries, was set up to identify and protect critical sites for shorebirds. They have a long way to go. For example, only 1 percent of the land around Boundary Bay, a site of global significance, is protected. And James Bay, a very crucial area, is threatened by Quebec's massive hydroelectric dams. But WHSRN is an important beginning that informs us of the global significance of the local battles.

Another illustration of the transnational behavior of wild species is the annual trip of monarch butterflies. To us denizens of the concrete-and-glass human habitat of downtown cities, the sudden appearance of a speck of color wafting along airstreams over busy streets is an unexpected delight. Often it is a monarch fattening up over the summer before a remarkable journey the length of the continent. Perhaps a hundred million monarchs east of the Rocky Mountains respond to their genetically programmed urges and navigate unerringly to their wintering grounds 3,000 meters (10,000 feet) above sea level in the mountains of central Mexico. At nine known sites, they gather to wait out the winter, garishly coating Oyamel fir trees like Christmas ornaments. In the spring, they turn around and begin the journey back. Most reach only Texas, where they lay their eggs on milkweed plants and die. Their offspring complete the trip to Canada.

The monarch's migration and transnational life cycle is one of nature's great spectacles, a global treasure. Public interest in the insects and the discovery of several of their wintering sites prompted Mexican president Carlos Salinas to establish butterfly preserves at five of the sites in 1986. After large numbers of monarchs died in the winter of 1992, Salinas announced a further commitment to protect them. His decree prohibits logging and agriculture in the heart of the sites and restricts other activities around them, but already two of the sites have been destroyed while the others have not been officially delineated.

All the known sites should be protected and their borders enlarged to include whole watersheds. Most of the land indicated in the decree is owned by *ejidos,* collective farms for local communities, which need an incentive to protect the butterflies. In order to make the preserves work, Mexico needs the will, money, and personnel.

As we pursue the economic benefits of the North American Free Trade Agreement (NAFTA), ecological consequences of "development" in the name of jobs and profit should not be ignored. If transnational biological treasures like monarch butterflies matter to us, NAFTA should include clauses that commit us to protect them by providing aid to Mexico to expand the number and perimeters of the butterfly preserves and to police them. We should also act in our own backyard, by, for example, removing milkweed, the food of monarchs, from Ontario's list of "noxious weeds."

For all people whose lives are enriched by appreciation of the beauty of monarch butterflies and their marvelous life cycle, national borders shouldn't deter our efforts to protect them.

If we "listen" to the land and its other inhabitants, we might have cities that conform to their geographic contours, farms that are divided by meandering ditches and dotted with ponds and sloughs, and parks that encompass entire watersheds. And perhaps, in some future time, war fought for some arbitrary piece of human turf will be recognized as the madness it is.

Update

Dr. Fred Urquhart of the University of Toronto discovered the monarch butterflies' wintering grounds in Mexico in 1975. Today Dr. Lincoln Brower is probably the world's foremost authority on them. He says the great butterfly migration is "an endangered biological phenomenon."

The insects must make their way through a barrage of chemicals on their way to Mexico from Canada through the United States. Manual mowing to clear roadsides, lanes for electric power lines, and agricultural fields has been replaced by herbicides, thereby depriving the butterflies of their food. As well, insecticides, which are heavily used, fail to distinguish between pests and desirable insects such as monarchs. Development of land for housing, roads, and farms also deprives the butterflies of important feeding areas on their way south.

In Mexico, the Oyamel fir forests that are crucial winter habitat have been steadily encroached upon by logging. About 2 percent of the fir forest is lost per year; at that rate, the forest will be gone by 2050. So fifteen years ago, Mexico decreed a monarch reserve of more than 16,000 hectares (62 square miles), prohibiting logging in five wintering sites. However, the people who lived within the reserve were not compensated for the loss of their land, and claiming poverty, they continued to log.

On November 9, 2000, Mexico tripled the protected areas to 56,000 hectares (216 square miles). But now, with an anonymous donation of $5 million to be built into a $30 million fund, local people are being compensated for the loss of their land, and new economic initiatives will provide alternatives to logging. The fund is jointly administered by the World Wildlife Fund and the Mexican Fund for Natural Conservation.

There's a Lot to Learn

AS POLITICIANS SCRAMBLE TO ESTABLISH A STRATEGY TO "GET OUR economy moving," science and technology are often cited as the key to economic success. But as long as politicians focus on glamor areas such as computers and biotechnology and demand quick returns for their investment, their strategies will fail. Today, the global biosphere is in trouble, and we are looking the wrong way in focusing on growth.

Politicians and government bureaucrats oversee natural resources—fish, trees, water, soil, air—on the erroneous assumption that we possess the knowledge and expertise to formulate sound "management" policy for these resources. The ever-increasing list of ecological crises—chemical spills, nuclear accidents, air pollution—makes it obvious that we lack enough hard data about the species makeup and their interaction in the planetary biosphere to act in an informed way.

Understandably, attention has been focused on tropical rain forests, which are the major repository of the world's species. But in North America, large areas of old-growth forest continue to be "liquidated" because experts boast they know how to "reforest" the land. It is a remarkable claim when you consider how ignorant we are. As insect specialist Tom Eisner once told me, he can go to Central Park in New York City any day and find a new insect species. That is in a big metropolis. Think how many undiscovered species there must be in old-growth forests.

Geoff Scudder, an entomologist and past head of the zoology department at the University of British Columbia, is a world expert on lygaeids, a group of insects that includes stink bugs. Not only is he familiar with virtually every known lygaeid species in the world, he has specimens from over four hundred genera that have never been identified! Scudder decries the lack of support for systematics, the science of collecting, naming, and classifying species, and says the need is as great in botany as it is in zoology. He recounted a field trip to the island of New Caledonia in the South Pacific: "Botanists in the Jeep with me suddenly leaped out when they spotted a flowering tree. They had been waiting for ten years because often positive classification can only be made with fruits and flowers. In one afternoon, they discovered six new genera of trees! And there are all kinds of trees that have never been seen in bloom."

Canadian foresters argue that since our forests lack the species diversity of their tropical counterparts, far more of the components are known. Yet the research platform recently built in a Sitka spruce tree in British Columbia's Carmanah Valley by the Western Canada Wilderness Committee was the world's first such platform built in a temperate rain forest. Already it is proving to be a research bonanza. University of Victoria graduate student Neville Winchester and his supervisor, Richard Ring, report that several new species and genera are among their collections from the platform. They estimate that a mere 30 percent of arthropods (insects, spiders, crustaceans, millipedes, and centipedes) in the area have been collected.

Winchester observes that many canopy insect species are flightless and sedentary, confined to highly specific microhabitats in old-growth trees. Clear-cut logging radically changes the landscape, thereby massively altering the flora and fauna and often eliminating many of the natural predators of insect pests of humans. Winchester says of the canopy-specific species, "The removal of this habitat will cause a decrease in biological diversity and initiate extinctions." His studies record "a unique fauna that is the product of long-term stability and is now habitat restricted," and he warns that clear-cutting of coastal old-growth "will reduce biodiversity and cause the permanent loss of these unique communities."

Whereas the importance of trees may be obvious to the layperson because of their potential medicinal properties or yield of fruit or wood, the value of identifying insects may not be. Yet because they are so numerous and varied, insects are extremely important components of ecosystems. Many flowering plants, trees, mammals, birds, and fish would not survive without them. As a geneticist who spent twenty-five years studying fruit flies, I can also tell you that the study of insects has provided some of our most fundamental insights into the mechanisms of human heredity, development, and behavior.

Based on his studies of tropical insects, the Smithsonian Institution's Terry Erwin estimated there may be 30 million insect species on Earth. Even if there are only five to 20 million species, the scale of our ignorance of planetary biodiversity is vast. And remember, identifying a species merely means giving a name to a dead specimen. It doesn't mean we have any basic information about its numbers, habitat, reproduction, behavior, or interactions.

Since we know so little about the biological world, we need more information about the extent of biological variability. But the basic information is trickling out too slowly, and one of the main reasons is scientific snobbery. Most financial support and publicity go to areas such as cancer research, computers, and biotechnology, or "sexy" fundamental subjects like astronomy and particle physics. Descriptive fields like taxonomy and systematics and even evolution and ecology are low in the scientific pecking order and therefore, not surprisingly, much more poorly funded.

This poor direction of funds makes no sense when ecosystems around the world are being devastated by human activity and species are becoming extinct faster than we are discovering them. Because of the low priority of systematics, it is difficult to recruit new students and those who do graduate in the area can't find jobs. Government positions for systematists in museums and research centers are being phased out, yet of those insect species already collected in Canada only half have been identified.

Any scientist should understand that our ignorance about the world is vastly greater than what we know. We construct policies for natural resources based on ludicrously skimpy and biased information. And because we prefer to study things that are large, microscopic organisms such as soil fungi, marine picoplankton, and even insects are low on the scientific status pole. We ought to be putting our effort into determining the extent and importance of this diversity, not destroying it before we have even found it.

Update

In 2002, scientists announced the elucidation of more than 98 percent of the sequence of three billion letters in the human genome. It was a stunning achievement that was accomplished years before expected. Whereas scientists trumpet the potential of the genome to reveal the causes and ultimately the treatment of hereditary diseases, to me the really exciting insight gained has been that genetic sequences exist within the human genome that are identical to sequences found in birds, insects, plants, and bacteria. The human genome reveals the depth of our relatedness to all other life-forms on Earth.

The challenge today is that our biological relatives—other species—are now vanishing from Earth at a rate estimated to be more than 130 a day. And we only "know" 10 to 15 percent of all species that exist. In other words, species are disappearing before we have even identified them. Yet to understand and emulate the productivity of Earth's biodiversity, we have to know what the component species are and how they interact.

That is why an exciting new megaproject has been proposed by the All Species Foundation: "The complete inventory of all species of life on Earth within the next 25 years—a human generation." The foundation proposes to describe and classify all remaining species on the planet to provide the knowledge base to assess the extent of environmental changes and devise the best conservation practices. The project, the foundation states, is the "key element in the maturing of ecology, including the grasp of ecosystem functioning and of evolutionary biology. It also offers an unsurpassable adventure: the exploration of a little-known planet."

The foundation is headed by an impressive list of scientists and conservationists, including E.O. Wilson, Peter Raven, and Terry Erwin, and green businessmen such as Paul Hawken and Stewart Brand. The plan is to raise tens of millions of dollars to support a global search for new species and to create a computer-based inventory of each one. Each species will be listed with every bit of available information about it. This is an initiative that is long overdue.

Elephants of the Sea

AS SPECIES VANISH WITH HORRIFYING SPEED, OUR ATTEMPTS TO protect dozens or even a few hundred that are rare or endangered seem far too little and too slow to counter the rate of loss. Attempts to breed a few dozen rare species in captivity cannot compensate for the tens of thousands disappearing annually. And even where a species has been rescued from oblivion, protecting its habitat often remains a struggle. So it gives a much-needed psychological lift whenever a good-news story is encountered.

This report is about one of the most fascinating creatures on Earth and suggests there is still great resilience and potential for recovery in the natural world—if we give it a chance. I found this story on a surf-battered beach on San Miguel Island in the Channel Islands National Park about 40 kilometers (25 miles) off the California coast and 140 kilometers (90 miles) north of Los Angeles. Even this close to a huge human population, there were hundreds of California sea lions and northern fur seals around me.

More sparsely interspersed were dozens of immense animals that looked like whales compared with the others. They were northern elephant seals, the first of tens of thousands that converge on the beaches by January. These massive animals may reach up to 5 meters (16 feet) in length, and a big male can weigh in at 2 tonnes. But the most striking feature of the species is an immense nose on the males that seems like a parody of a tapir's proboscis. When a male reared up aggressively, it towered above our heads and inflated its nose as an awesome threat display. Letting out a barrage of bellows, hisses, belches, and honks, males and females give plenty of warning to stay away from their sharp teeth.

Scientists Brent Stewart and Bob Delong filled me in on some of the incredible features of elephant seal biology. They track the animals with radio transmitters, while high-tech microprocessors glued to the seals' skin record depth, light, and temperature at minute intervals for four months.

Since the animals return to San Miguel to shed their skin, the researchers have a high rate of recovery for their recording instruments. What they find is truly mind-boggling. The males migrate from Southern California to Alaska—that's a 10,000-kilometer (6,000-mile) round trip—twice a year!

Even more astonishing is the pattern of diving that is revealed by the recorders. Elephant seals can stay submerged for longer than an hour. They dive for an average of twenty-two minutes to depths exceeding 1,500 meters (5,000 feet) and spend a quarter of their dive foraging for squid.

There is no surface light at such great depths, but bioluminescence of the deep-sea creatures themselves can be captured by the immense eyes and super-sensitive retinas of the seals. After surfacing from a deep dive, an elephant seal reinflates its lungs for a couple of minutes, then dives again, repeating this action around the clock for weeks on end.

I have barely begun to relate the astonishing traits of these fascinating animals. Now here is what makes this a happy story. Immense middens on San Miguel reveal that aboriginal people hunted elephant seals in large numbers as far back as 11,000 years ago. No one knows how extensive the original elephant seal populations were, and the only guess I have seen advanced is perhaps 250,000.

What we do know is that they were easy prey for the harpoons and guns of Europeans, who soon recognized their worth as sources of oil. Whalers slaughtered them and simply mixed elephant seal oil with whale blubber. By 1869, northern elephant seals were "as good as extinct" for commercial purposes. Their rarity merely stimulated collecting expeditions from museums that wanted specimens before the species vanished. In 1891, a party sent by the Smithsonian Institution discovered eight elephant seals at Guadalupe off the Baja and promptly killed seven of them in the full knowledge that they might be among the last of their kind on Earth. Again, no one knows how low the population got, but estimates range from less than a hundred to a few hundred.

Now comes the remarkable part. Shielded by their relative inaccessibility, by wildlife protection laws, and by the fact that they don't compete directly for prey with humans, the northern elephant seal began a spectacular comeback. Today, their population stands at 130,000, a number some suggest could be approaching their original levels. And it is not just elephant seals. Northern fur seals were so heavily hunted that by 1835 they were considered extinct. Then, in 1969, when Bob Delong came to San Miguel to do research for his Ph.D., he discovered 100 females, 36 pups, and one mature male. Their numbers, too, have exploded. Last year, 1,000 elephant seals and more than 1,000 harbor seals were born on San Miguel.

It is inspiring to realize that there, in the most heavily populated part of North America, where the waters are dotted with offshore oil platforms as well as commercial and sport fishing boats, pleasure boats, and transport ships, the ocean remains rich enough to support these expanding populations of impressive marine mammals. Perhaps we can share the planet with our biological relatives.

I hope so.

The Thrill of Seeing Ants for What They Are

WATCH A CHILD'S FACE WHILE SHE'S POKING AROUND IN A TIDAL pool, peering under a rotten log, or netting tadpoles in a pond. The joy of surprise and discovery is instantly visible. When my daughter Severn was three or four, she asked me how worms are able to move through the ground. I told her they simply eat soil at the front end and poo it out the back. "Oh," she responded, "so dirt is worm-food?" And I could see a smile of delight at this new thought.

Charles Darwin shared Severn's fascination with worms. When he heard someone had estimated that there were 53,767 worms per acre, Darwin is reported to have "figured the worm population swallowed and brought up ten tons of earth each year on each acre of land. Earthworms therefore were ... constantly turning it inside out. They were burying old Roman ruins. They were causing the monoliths of Stonehenge to subside and topple." Darwin concluded: "Worms have played a more important part in the history of the world than most persons would at first suppose."

One of the pleasures of science journalism is learning something that clicks on a light, suddenly providing an entirely new perspective on things around us. For example, in the 1970s, the biologist Lynn Margulis suggested that structures called organelles, found within cells of complex organisms, are actually the evolutionary remnants of bacterial parasites.

She pointed out that organelles are able to reproduce within a cell and even possess DNA and distinct hereditary traits. So, Margulis proposed, organelles were once free-living organisms that invaded cells and were eventually integrated into the host. In giving up their independence, these bacterial relics received nourishment and protection from the host cell. Looking in the mirror now, I see a reflection of a community of organisms inhabiting trillions of cells aggregated as me.

In an interview for *The Nature of Things,* I once asked Harvard University's eminent biologist Edward O. Wilson why ants are so successful. His entire career has been spent studying these ubiquitous insects, and he became animated with enthusiasm as he answered:

> There are only a few tens of thousands of species of ants, compared with millions of other nonsocial insects, but they dominate the world. Their secret is superorganism. A colony of ants is more than just an aggregate of insects that are living together. One ant is no ant. Two ants and you begin to get something entirely new. Put a million together with the workers divided into different castes, each doing a different function—cutting the leaves, looking after the queen, taking care of the young, digging the nest out, and so on—and you've got an organism weighing about 10 kilograms [22 pounds], about the size of a dog, and dominating an area the size of a house.
>
> The nest involves moving about 40,000 pounds [18,000 kg] of soil and sends out great columns of workers like the pseudopods of an amoeba, reaching out and gathering leaves and so on. This is a very potent entity. It can protect itself against predators. It can control the environment, the climate of the nest. When I encounter one of these big nests of leaf-cutter ants, I step back and let my eyes go slightly out of focus. And what you see then is this giant, amoeboid creature in front of you.

It was a thrilling description that made me think about ants in a very different way.

In 1992, scientists in Michigan made the astounding announcement that the network of mycelia, threadlike extensions of fungi found in the ground, could be derived from one individual, not an aggregate of different organisms. They reported a single organism that extended throughout

16 hectares (40 acres). Not long after, biologists in Washington reported a fungus covering 607 hectares (1,500 acres).

This fall I was filming in a grove of quaking aspen, the lovely white-barked trees whose leaves shimmer at the slightest puff of air. I learned that what I thought was a group of individual trees was, in fact, a single organism. Like a strawberry plant that can spread asexually by sending out runners that grow roots and sprout leaves, quaking aspen multiply vegetatively. From one tree, shoots may grow up from a root 30 meters (100 feet) away.

The aspen is another kind of superorganism that can exploit a diverse landscape—parts of the organism may grow in moist soil and share the water with other portions perhaps growing in mineral-rich soil higher up. In Utah, a single aspen plant made up of 47,000 tree trunks was discovered. It covered an area of 43 hectares (106 acres) and was estimated to weigh almost 6 million kilograms (13 million pounds).

We know so little about the varieties of species that exist and the specific attributes of individual organisms that we have endless opportunities for discovery. It is an exciting prospect to anticipate the delicious surprises that await us.

The Case for Keeping Wild Tigers

CONSERVATIONISTS PREDICT THAT MEMBERS OF ONE OF EARTH'S most magnificent species, India's Bengal tiger, may disappear completely from the wild within five years. The tiger is high on the food chain, so its disappearance would not be as ecologically disruptive as the loss of insects or fungi, for example. Furthermore, tigers require a lot of space to move in search of prey and mates, therefore conflicting with growing human populations and demands. So one may wonder what it matters if one more species, especially a large carnivore that has long been feared as a "man-eater," disappears.

If the tiger goes from the wild, we will suffer spiritually. The tiger is part of the rich mythology and folklore of humankind, a symbol of our love-fear relationship with nature. Its beauty, size, ferocity, and power inspire awe and terror simultaneously, reminding us of our own frailty and vulnerability. Like the beluga, marbled murrelet, and spotted owl, the tiger is an "indicator" species whose fate will tell us whether we are capable of sharing the planet with others.

In preparing programs on wildlife, I am constantly shocked at the speed and scale of loss. Years ago, I stood in the Delta Marsh south of Lake Winnipeg during the fall migration of waterfowl. The sky was filled with a cacophony of calls and dozens of chevrons of geese at one time. It was thrilling to know that wild creatures still act out their genetic destiny as they have for millennia.

But those waterfowl migrating up the central North American flyway were a mere vestige of the past. A century ago, I was told, the birds literally filled the sky from horizon to horizon on their great flights along the continent. I was also amazed to learn that grizzly bears were not always confined to West Coast mountains—they once populated foothills and prairie grasslands all the way to Ontario and down to Texas and California. They were prairie animals living on the great herds of bison, which were also exterminated.

Elders who have lived their entire lives along the East and West Coasts, across the Prairies, or above the Arctic Circle recount depressingly parallel stories of decline in abundance and variety of living things. Their memories encompass a sweep of time that is often lost as time shrinks and the pace of life speeds up.

Farley Mowat's classic book *Sea of Slaughter* is a lament for the decimation of walruses, whales, and seabirds, and the confinement of the survivors to a fraction of their former range. Mowat's book anticipated the collapse of the northern cod, which has been catastrophic for Newfoundlanders.

In cities, gleaming displays of vegetables, fruit, and meat in supermarkets create an illusion that the Earth's abundance is endless. It's easy for urban dwellers to believe this fantasy of a world without limits when we are immersed in a human-created landscape and few of us get to experience seasonal rites of nature anymore. But because our reference point is the urban setting and what we experience, we don't see the impoverishment of nature over the past two centuries.

Imagine what uplifting opportunities we have lost. It boggles the mind to think of the sight of tens of millions of bison moving across the Prairies. The plant intake and fecal output of those giant herbivores created the prairie grasslands, supporting the countless plants, insects, fungi, and birds that evolved along with them.

And passenger pigeons! Eyewitness accounts tell of day after day of darkness as billions of birds flew by. It is said that forests rang with the cracking of branches collapsing under the weight of the resting birds. Night hunters, who slaughtered them to ship to Europe, slithered on guano-coated forest floors. The elimination of passenger pigeons in a mere blink of evolutionary time must have had catastrophic ecological reverberations.

In our total preoccupation with exploitation and consumption, we have become impoverished spiritually. The ultimate source of our solace, inspiration, companionship, and sense of place has always been nature. As we tear at the web of living things of which we are a part, we not only threaten our own biophysical needs, but also eliminate our evolutionary relatives and fellow Earth beings. We live in a terribly degraded and empty world in which the companionship of our evolutionary kin is replaced by clever toys from the industrial juggernaut to keep us distracted.

The fact that my grandchildren will grow up in a world in which tigers exist only in zoos, books, and videos pains my soul.

How Little We Know

THE TWENTIETH CENTURY HAS OFTEN BEEN CALLED THE AGE OF science. It is true that the numbers, technological dexterity, and knowledge base of scientists have expanded tremendously. Our ability to create new drugs and chemicals, travel into outer space, and explore the human body has grown explosively. Yet it is one of those seeming contradictions that while our knowledge and technological capacity grow exponentially, our basic descriptive knowledge of the Earth's creatures remains abysmally small.

A visit to the Monteverde Cloud Forest Preserve in Costa Rica gave me an opportunity to reflect on nature and the role of science in understanding it. In a tropical rain forest, one's senses are assaulted by air laden with humidity and mysterious, earthy smells, a steady drone of insects punctuated by screams and songs of birds, and, everywhere, a dazzling profusion of plant life.

In temperate regions, forests are made up of vast numbers of relatively few plant and animal species, but in the tropics, individual trees support a veritable community of diverse animals and plants. The variety of life in a tropical forest makes it difficult even to know where to look or what to look for. It's best just to stand or sit in one place and focus on one sense at a time. With luck, a spectacular quetzal bird will land nearby or a chattering band of white-faced monkeys will lope across the canopy. Gradually, one becomes aware of flowers of all sizes, shapes, and colors, and around them, insects. Most people focus on birds and mammals, but insects are the most abundant and diverse group of organisms in the forest. For every human being on the planet, there are at least 200 million insects, and they are endlessly fascinating.

"A person with a Ph.D. could spend a lifetime studying one square meter of the forest floor," I commented to a guide. "Yes," he replied, "and the square meter right next to it could be completely different."

Unlike species in temperate forests, which may be found ranging over hundreds of kilometers, species in a tropical rain forest may be confined within a few hundred meters. That's why destruction of tropical rain forests causes so much species extinction.

A powerful although probably apocryphal story tells of a student who asked to work for a Ph.D. in the lab of a great German zoologist. The professor handed the student a common minnow and told him to come back when he had learned everything about it. The student measured the fish, dissected it, inspected its scales, and a week later came back with a notebook filled with information.

"Go and study it some more," the student was told. So he collected other specimens, looked at their internal organs, checked reproductive cells under the microscope, and six months later returned with several notebooks full of material.

"You still have more to learn," the professor said, sending him away once again. The student realized that he hadn't studied the animal in its natural habitat, what its predators were, or how it behaved with its own kind. He went away to do more studies—and never returned.

Today, the volume of research in publications has grown so rapidly that it is impossible for scientists to be well informed in all areas. For example, I graduated in 1961 as a "geneticist," specializing in a subdiscipline of biology. In the 1990s, my expertise would be described as the study of chromosome behavior in the fruit fly. As scientists penetrate deeper into the mysteries of nature, the breadth of their knowledge becomes more limited.

In 1989, I visited the research station of the World Wildlife Fund in the Amazon rain forest near Manaus, Brazil. There were three frog experts in the camp, and their knowledge and skills were impressive. At night in pitch-dark, they could find frogs a centimeter (½ inch) long. But when I asked one about a bird and a strange plant we saw, he replied, "Don't ask me. I'm a herpetologist."

Later that year, I stayed in the traditional Kaiapo Indian village of Aucre, deep in the Amazon rain forest. Each time I asked a Kaiapo about a plant or animal, he or she identified it by name and told an anecdote about it. Kaiapo knowledge of forest flora and fauna is by no means complete, but it has enabled the people to survive in harmony with the forest biodiversity. Science, in contrast, allows us to extract great detail by sacrificing that sense of nature's vast breadth and immeasurable complexity. In the Monteverde Cloud Forest, one is overwhelmed by the immensity of our ignorance, a sense of humility about our abilities, and a reverence for nature, which put our sense of achievement into perspective.

So how much do we know of life's diversity? Of the vast range of life-forms on this planet, scientists have identified perhaps 1.4 million species. That includes about 750,000 insects, 250,000 plants, and 47,000 vertebrates. The rest are invertebrates and microorganisms. Of the vertebrates, 4,300 are mammals, 9,000 birds, 4,000 reptiles, 3,500 amphibians, and the rest fishes. That gives you an idea of the spectrum of life-forms on Earth.

Today, loss of biodiversity—that is, species variety—is considered by biologists to be a global crisis second only to the threat of nuclear war. Most of the extinction is taking place in tropical rain forests that occupy a mere 6 percent of the land's surface but contain at least 50 percent (some suggest 60 to 80 percent) of all species on Earth.

Most tropical species are found in the foliage of the forest canopy, which poses a formidable challenge to scientists, since it is a layer of biomass suspended 30 to 50 meters (100 to 160 feet) above the ground. It is impossible for a scientist to move around collecting and observing there, meaning that we can only catch glimpses into this ecosystem. In one study, a blast of insecticide was shot into the canopy of a Peruvian rain forest and almost every species of insect recovered had never been seen before. That study led to the estimated number of species put globally at 30 million.

It is difficult to comprehend the scale of variety in tropical rain forests. In 1 hectare (2.5 acres) of a Peruvian Amazon forest, 41,000 species were identified. In the peninsula of Malaya alone there are 2,500 tree species—that compares with only 700 in all of the continental United States and Canada. Harvard biologist E.O. Wilson reports: "From a single leguminous tree in the Tambopata reserve of Peru, I recently recovered 43 species of ants belonging to 26 genera, about equal to the entire ant fauna of the British Isles."

Even with the little knowledge we have of tropical forests, it is clear they play important planetary roles. The forests absorb massive amounts of water and liberate it slowly, thereby ameliorating the cycle of water in weather and climate.

In the late 1970s, the United Nations Food and Agriculture Organization (FAO) estimated that 40,000 square kilometers (15,000 square miles) of tropical forest were disappearing annually, a rate that would destroy them all within sixty years. The FAO report assumed no increase in human population or demand on the forest and was far too optimistic. In the past thirty years, world consumption of tropical forest products has increased 1,500 percent and is still accelerating.

A hundred years ago there were 15 million square kilometers (6 million square miles) of tropical forest. Today, that number is down to 9 million (3.5 million square miles). Present destruction is about 200,000 square kilometers (77,000 square miles) annually (an area the size of South Dakota or Newfoundland), taking with it an estimated number of species recently revised from 20,000 to 50,000.

Tropical forests are not cleared only for local agriculture. They are also destroyed to bring in foreign revenue. Much of the destruction of forests in Mexico and Central America is for land to grow cattle, which are sold to hamburger chains in North America. It is estimated that raising cattle in the South saves companies five cents on every patty.

Countries that have exported their wood are running out. Today, 95 percent of Madagascar's rich and unique rain forest is already gone. Ghana and the Ivory Coast now ban export of several previously commercial tree species, and Nigeria has stopped exporting altogether.

The demand of the industrialized countries for a constantly expanding economy and for more consumer goods is the direct cause of the current spasm of species extinction that has no counterpart since the dinosaurs disappeared.

Update

Two centuries ago, rain forests covered 14 percent of the land; today, they cover less than 6 percent, and at the current rate of loss, they will all be gone in less than forty years.

Rain forests are the richest, oldest, most productive and complex ecosystems on Earth. Biologist Norman Myers says, "Rain forests are the finest celebration ever known on the planet." In 1989, Myers estimated the rate of destruction of rain forests was about a hectare (2.5 acres) a second. That is, every second, an area of forest equal to two football fields was being logged, burned, or flooded. That's 60 hectares a minute, 86,000 a day, and 31 million a year.

In his book *The Diversity of Life,* Edward O. Wilson estimates that 137 species go extinct a day. That's 50,000 a year, and much of it from destruction of rain forest. Humans are not immune to the consequences of rain forest destruction. Conservation writer Catherine Caulfield says Brazil's rain forests supported 6 to 9 million indigenous people in 1500. By 1992, as Brazil's forests were falling at the rate of 2.25 million hectares (5.5 million acres) a year, there were fewer than 200,000 indigenous people left.

The loss of tropical rain forests I wrote about in the 1980s continues with catastrophic speed as a direct result of logging, cattle ranching, mining, oil extraction, hydroelectric dams, subsistence farming, and fuelwood gathering, and indirectly by overconsumption, debt burden, overpopulation, and unbridled development. It is such a shortsighted disaster. It is estimated that the economic returns on a hectare (2.5 acres) of Peruvian rain forest would be $6,820 annually for fruits, latex, and timber, $1,000 if clear-cut, and $148 for cattle pasture.

The Amazon rain forest is so immense that even with all the depradations, more than 70 percent remains intact. There is still time to save most of the biggest tropical rain forest left on the planet.

The Power of Diversity

IN PAST DECADES, THE SCIENTIFIC COMMUNITY HAS UNDERGONE A tremendous expansion and knowledge has increased proportionately. However, too often the accumulation of information is mistaken for knowledge that provides understanding and control. We can't afford to make such an assumption, because it fosters the terrible illusion that we can "manage" wilderness and it has resulted in destructive consequences.

Globally, old-growth forests are being cleared with alarming speed. In the past decades, geneticists have made a surprising finding that foresters should heed. When seemingly homogeneous populations of organisms were analyzed using molecular techniques, they were unexpectedly found to be highly diverse. When looked at from individual to individual, the products of a single gene are found to vary considerably. Geneticists call such variability *genetic polymorphism,* and we now know that a characteristic of wild populations of any species is a high degree of genetic polymorphism. Apparently, maximum genetic diversity optimizes the chances that a species can withstand changes in the environment.

When individuals of the same species are compared, their patterns of genetic polymorphism differ from region to region. Thus, whether a tree, fish, or bird, different geographic subgroups exhibit different spectra of variation. So Ronald Reagan was dead wrong—if you see one redwood tree, you *haven't* seen them all. Stanford ecologist Paul Ehrlich says, "The loss of genetically distinct populations *within* species is, at the moment, at least as important a problem as the loss of entire species."

The biological value of diversity can also be applied to a collection of species. A forest is more than an assemblage of trees; it is a community of plants, animals, and soil microorganisms that have evolved together. This aggregate of species creates a highly resilient forest with a great capacity to recover from fire, flooding, landslides, disease, selective logging, or storm blowdowns. That's because the diverse species remaining in the surrounding areas can replenish the damaged parts. Clearly, we should try to maximize forest diversity by protecting as many old-growth forests as possible. That's the best way to ensure the maintenance of a broad genetic base on which the future of the forestry industry will depend.

There is a way to illustrate the power of diversity by looking at our bodies. Just as a forest is made up of vast numbers of individuals of different species, *we* are an aggregate of some 100 *trillion* cells that vary in size, shape, and function. These different cells are organized at many levels into tissues and organs that all come together in a single integrated whole—a functioning body.

The collective entity that is each of us thus is a mosaic of an immense array of different cell, tissue, and organ types that have enormous resilience and recuperative powers. If we suffer a cut, bruise, or infection, the body has built-in mechanisms to overcome the assault. We even have the ability to regenerate skin, liver, blood, and other body parts and compensate for damage to the brain and circulatory systems. We can function pretty well with the loss of some body parts, such as a digit, tonsils, or teeth. In short, our bodies can absorb considerable trauma and recover well, a tribute to cellular diversity in form and function.

If we amputate large parts of the body, we can still function and survive. Thus, we can live with the loss of limbs, eyes, ears, and other parts, but each loss confers greater dependence on other people and on human technological ingenuity to compensate for lost abilities. With the power of modern science and high technology, we can make artificial substitutes for teeth, bones, skin, and blood, and we have even devised machines to take over for the heart, lungs, and kidney. In principle, it should be possible for an individual to survive the combined loss of organs that are not absolutely necessary for life and those that can be mimicked by machines. Thus, a blind, deaf, quadruple amputee who is hooked to a heart-lung and kidney machine could live and would still be a person, but one with capabilities and resilience radically restricted in comparison with a whole individual. Essentially, such a patient would be a different kind of human being, created by and dependent upon human expertise and technology.

In the same way, a forest bereft of its vast biodiversity and replaced by a limited number of selected species is nothing like the original community. It is an *artifact* created by human beings who foster a grotesque concept of what a forest is. We know very little about the basic biology of a forest community, yet road building, clear-cut logging, slash burning, pesticide and herbicide spraying, even artificial fertilization have become parts of silviculture practice. The integrity of the diverse community of species is totally altered by such practices with unexpected consequences—loss of topsoil, death from acidification, weed overgrowth, disease outbreak, insect infestation, and so on. But now, caught up in the mistaken notion that we have enough knowledge about forests to "manage" them in perpetuity, we end up ricocheting from one contrived Band-Aid solution to another. Medical doctors today are struggling to readjust their perspective to treat a patient as a whole individual rather than as an aggregate of autonomous organ systems and ecopsychology recognizes the relevance of our surroundings on our psychological health. A similar perspective has to be gained on forests. The key to development of sustainable forests must reside in the maintenance of maximum genetic diversity both within a species and between the species within an ecosystem. If we begin from this basic assumption, then the current outlook and practices in forestry and logging have to be radically overhauled.

Owning Up to Our Ignorance

AT THE BEGINNING OF AUGUST 1987, MORE THAN TWO THOUSAND people hiked over a mountain pass to camp at the head of the Stein River Valley in British Columbia. They were there for a festival celebrating another of this planet's special places.

At the Stein Festival, I met a forester who had accused me on radio of being "too emotional" and having "little factual information" to back up my support for the preservation of the Stein. (I have never figured out what is wrong with being emotional about something that matters—I *am* emotional and I do not apologize for it.) But when I challenged his claim that foresters know enough to replace the likes of the trees they would cut down, he replied, "We're almost there. We've learned a lot and we'll soon know everything we need." He was confident his children could look forward to logging the kind of trees he wanted to harvest in the Stein.

Scientists *have* learned a lot in the past half-century. Many of their insights are truly mind-boggling—models of atoms and subatomic particles, black holes, DNA, and the immune system. These investigations have been accompanied by the invention of technological tools to disrupt and alter much of the natural world. Our manipulations are often extremely powerful and yield immediate results, yet they are crude approximations of what exists in nature. A mechanical heart is quite a technical achievement, but this simple pump does not come anywhere near duplicating the real thing. Unfortunately, people like that forester have equated our power to effect short-term changes in nature with long-term control and progress. This is sheer arrogance.

Canadian support of science has always lagged behind that of other industrialized countries, and forestry, not being a high-profile area, has always suffered a severe shortage of funds and top students. How can we believe we comprehend the diversity of organisms and the complexity of their interactions when our analytical techniques and expertise are so limited?

I spent twenty-five years of my life practicing science. My area of genetics has been a high-profile field that has attracted top scholars and funds for decades. My entire career was spent focused on one of the estimated thirty million species on this planet—a common fruit fly. This fruit fly has been studied by geneticists since 1909 and for over seventy years has been at the center in the study of heredity. Tens of billions of dollars have been invested to pay the salaries of thousands of geneticists who have devoted their lifetimes to studying the fruit fly. Four scientists who studied fruit flies have already won the Nobel Prize, and no doubt there will be more in the near future.

The results of all this effort have been impressive. It is possible to create all kinds of mutant strains affecting the fly's behavior, anatomy, and viability. We can take a single cell from a fly and clone it to create another fly. We can extract pieces of DNA, alter them, and reinsert them into flies. We can make flies with wings growing out of their eyes, legs growing out of their mouths or antennae, four wings instead of two, twelve legs instead of six. We have gained tremendous insights and manipulative powers.

But you know something? After using hundreds of thousands of person-years of research, billions of dollars of grant money, and all the latest equipment to study this one species, we still don't know how common fruit flies survive the winter. We still can't understand how a fly's egg is transformed first into a larva, then a pupa, and then an adult, something every fly manages easily. There is a species of fruit fly so closely related to the one I've studied that only a handful of specialists in the world can tell them apart, but the flies have no problem. It turns out that we have very little understanding of the basic biology of this one species, and it is only one of tens of thousands of species of fruit flies! If we know so little after all that, how can we possibly think that we have accumulated enough knowledge to enable us to manage complex ecosystems like the Stein Valley? As a scientist, I have been overwhelmed not with the power of our knowledge but with *how little we know.*

We have become so intoxicated with our clever experiments and increasing knowledge that we forget to see the intricate interactions and nearly infinite complexity that exist in the living world. Instead, smug in our faith in the knowledge of scientists, we perpetuate the notion that we already know enough to cut down the last areas of wilderness. Those "scientists" and "experts" who speak so confidently about the logging industry's ability to mimic the virgin stands that are being cut down reveal that they literally cannot see the forest for the trees.

Update

In the late 1980s, the Lytton Indian band had begun to hold annual Stein Valley Festivals to celebrate the place called Stagyn, meaning "hidden place," by the First Nations. Participants included musicians such as Gordon Lightfoot, John Denver, Buffy Sainte-Marie, and Bruce Cockburn, and attendance grew to tens of thousands. In 1995, the British Columbia government established Stein Valley Nlaka'pamux Provincial Park, jointly managed by the Lytton First Nation and B.C. Parks. Encompassing 107,191 hectares (almost 265,000 acres), the magical place would be protected for future generations in perpetuity.

In the mid-1990s, the David Suzuki Foundation began to establish relationships with the First Nations communities in remote areas along British Columbia's north and central coasts, including Haida Gwaii. Together, the foundation and the First Nations formed Turning Point, an organization of First Nations working together and supporting each other to protect their traditional lands. In April 2001, the B.C. government signed a historic document with the Turning Point First Nations, entering into agreements to negotiate the future of the land. Forest companies, environmental groups, ecotourism companies, and local municipalities added their support. The David Suzuki Foundation, at the request of Turning Point, put together a document on ecoforestry called "A Cut Above," which outlined nine critical principles underlying sustainable use of forests. It is now time to see whether all the goodwill and support will result in First Nations control over their lands and their use in an ecologically sustainable way.

A Walk in a Rain Forest

TO MANY PEOPLE, THE NAME COLOMBIA CONJURES UP IMAGES OF coffee or drugs. But to biologists, Colombia is home to one of the richest ecosystems on the planet, the Choco tropical rain forest, pinched between the Pacific Ocean and the Andes mountain range. It extends from Panama through Colombia to Peru.

Chugging from Bahia Solano to Utria National Park on the *Jestiven,* a wooden boat, I am accompanied by Françis Hallé, a French expert on tropical forests. Hallé is famous for having created a huge pneumatic platform that can be erected on the forest canopy, where researchers can explore 600 to 800 square meters (700 to 950 square yards) of the treetops.

Hallé points out the thick cloak of trees extending to the waterline. "The first thing people do when they invade such a virgin forest," he says, "is to clear the trees along the shore." Despite the difference in vegetation, the tree-covered mountains and pristine bays remind me of British Columbia.

Utria National Park was formed in 1987 and covers 54,300 hectares (134,000 acres) of spectacular forest. In a heavy rain, I set off alone to walk across a peninsular saddle along a thin path that is a slimy ribbon of red mud. Serpentine tree roots coil along the forest floor to suck nutrients from the thin topsoil and anchor the immense trunks in place. Though impediments on level ground, the roots provide welcome hand- and footholds on the steep hills.

In the forest, temperature and light intensity immediately drop. Thirty meters (100 feet) overhead, the canopy blocks out the sky, preventing growth of the heavy underbrush we think of as jungle. The steady rainfall is intercepted by foliage, so the water doesn't pound onto the soil. Even though it has rained constantly, the water in the creeks is crystal clear.

The ground is littered with leaves. In temperate regions, we classify trees as deciduous or evergreen, but here the trees shed leaves year round. Instead of building up to form thick humus, however, the leaves quickly become food for insects and fungi and thus are recycled back up into the forest biomass.

It's easy to walk along creek beds or through the trees with little vegetation to hamper movement. The noise is constant, a cacophony of the buzzing, clicking, and humming of insects and frogs. Walking quietly and slowly, eyes adapting to the shadows and shapes, one begins to notice movement that betrays a frog, a butterfly, a bird. A cosmos of complexity opens up.

Back on the boat, Hallé informs me that "jungle" is a word from India referring to the tangle of secondary growth that results after the initial forest is cleared. It is an insult to call a primary forest a jungle, he says. He draws my attention to trees with special properties—the hard white "tagwa" seeds, six to a cluster within an armored shell, that can be carved like ivory; fruit trees; parasitic air-breathing plants, lianes, orchids. But when I bring a seed or leaf, he often admits he has no idea what it is. When I ask how much taxonomists know of the species residing in tropical rain forests, Hallé makes a gesture of futility and replies: "It's an impossible mess." He tells me individuals of one species are usually spaced far apart, and each may house different spectra of associated species. That's the reason our ignorance is so vast.

Hallé believes the fabled diversity within a tropical rain forest gives it its stability. When one or a few trees are removed, the opening in the canopy allows light to reach the forest floor and stimulates a succession of plants. Over time, like a small nick in the skin, the opening is healed and filled in. But remove a large section of trees and, like a mortal wound, this large gap in the forest cannot repair itself.

In the rain forest, a destructive parasite is readily controlled because its target species is not concentrated in an area the way species are in temperate forests. "There's no need for pesticides," Hallé tells me, "because the forest is too diverse to allow an outbreak." Similarly, an introduced exotic species can't explode as rabbits did in Australia or purple loosestrife in Canada, because there are too many predators able to attack them. So biodiversity is not just a descriptive property of tropical rain forests, it is the very mechanism of its stability for survival.

World demand for lumber and pulp continues to rise, but forest plantations cannot deliver wood of quality or quantity. That's why deforestation continues to claim the great forests of the planet and threatens the Choco.

The Choco is the traditional home of perhaps thirty thousand aboriginal people belonging to three main groups—Embera, Waunana, and Cuna—who continue to live as they have for thousands of years, depending on the forest for their food, medicines, and materials.

From the airport at Bahia Solano, we take a bus up the coast to the village of El Valle, which is populated by descendants of African slaves who were brought to mine gold more than four hundred years ago. We rent a dugout with a motor and guide to take us up the Boro Boro River. After about three hours, we finally leave the plantations, cleared fields, sugar cane, and breadfruit trees to enter primary rain forest. As the river narrows, we drag the dugouts across shallow riffles and around fallen trees and logjams. At one point, we unload the boat and sink it to push it under a huge log blocking the river.

Night falls early and quickly in the tropics, and as the light fades, we know we are still hours away from our destination, the Embera village of Boro Boro at the junction with the Mutata River. Five hours after nightfall, we finally reach the settlement, exhausted, wet, but exhilarated by the adventure. Hammocks and mosquito nets are slung in the tiny school, and we soon accompany the frog calls with snores.

Boro Boro is home to eighty-four people living under thatched huts built on supports 2 meters (6.5 feet) above the ground. The tiny cluster of buildings is surrounded by small fields of domesticated plants. Life here revolves around the river for bathing, laundry, food, and transportation. A three-hour hike up the Mutata ends at spectacular falls that drop 400 meters (1300 feet) into a huge pool that is considered the source of life and power in the river. The people of Boro Boro fear the power of the place and stay away. Only the shaman goes to the pool to perform rituals that will ensure the fecundity of the river and forest.

The villagers tell us they want to keep their culture and way of life. They have heard of proposals to develop the area, which one Colombian prime minister referred to as the country's "piggy bank." The Pan American Highway, nearly finished, was stopped only when the minister of the newly formed environment ministry threatened to resign if it weren't. There are other proposals to build superports on the coast, a network of highways to link the ports to cities, and huge dams to deliver electricity to isolated villages. The familiar notion of "development" by extracting the resources of the forest is irresistible in Colombia too.

Colombia's forests, of which Choco is an important part, have the greatest number of known bird species and orchids (19.4 percent of all the world's known bird species, compared with 17.6 percent in Brazil and 15 percent in Africa), the second greatest number of amphibians, the third greatest number of reptiles, and one of every five bats. This rich tapestry of living things is beyond any scientific comprehension and, if destroyed, will never be duplicated or re-created.

There are people who have had the knowledge and expertise to make a living from these forests for millennia, but their futures are as uncertain as the fragile ecosystems that are their homes. The 1987 United Nations report *Our Common Future* stated: "It is a terrible irony that as formal development reaches more deeply into rain forests, deserts and other isolated environments, it tends to destroy the only cultures that have proved able to thrive in these environments."

Indigenous people throughout Colombia are organizing to resist incursions into their land. In the Choco, OREWA (Organización indígena regiònal Embera Wounaan) was formed to represent the indigenous groups Embera, Waunana, and Cuna. But in the government discussions about the future of the Choco, the indigenous people who have always occupied the forests are seldom involved.

The predicament is complicated by an Afro-Colombian population that outnumbers the aboriginal people by ten to one. After escaping slavery, the blacks were able to survive in coastal villages for two hundred to three hundred years. Lacking the indigenous culture and knowledge base built around the forest, the blacks have eked out a living and are desperate for the material benefits of modern life.

In negotiations with the government, OREWA has included Afro-Colombians as stakeholders in the forest lands. But impoverished people are easy prey to the blandishments of developers. Promises of jobs, electricity, and television tempt them to welcome roads and ports. To them, the forest is a resource that can be converted to money. If we in the rich nations of the world haven't been able to resist the siren's call of development, why should people who start out with far less?

Environmentalists in industrial nations of the North are concerned about the fate of tropical rain forests, which have been labeled the "lungs of the planet" and the "wellsprings of biodiversity." In Colombia, Latin Americans demand to know why they are expected to save the forests when countries in the North haven't protected theirs. In the debate about vanishing forests, the people who live in them are often forgotten.

Traveling through the Choco rain forest along mud tracks, one can't help but wonder why magnificent forests like this are being traded for squalid towns and villages of impoverished people and scrawny cattle grazing on barren hills. Is there no other way to create income for the human residents while preserving the forest ecosystem?

According to Françis Hallé, there is. He has spent his life studying plant growth in the canopy of tropical rain forests. When I ask him whether we know enough to cut down the likes of the Choco and regrow it, he replies, "Absolutely not!" He points out that a tree plantation is not a forest and that rapidly growing species like eucalyptus or pine imported from other parts of the world seldom perform as expected. Hallé says ideas developed from northern temperate forests are inappropriate for the tropics, where vegetation and soil are completely different.

Throughout tropical countries of Africa, South America, and southeast Asia, Hallé finds a sophisticated human practice called agroforestry that has sustained communities for hundreds, if not thousands, of years. Hallé has observed carvings on Indonesian temples depicting agroforestry practices about A.D. 1000.

Agroforestry requires a profound knowledge of plants that can be used for a variety of needs. Useful plants are collected from intact primary forests and deliberately planted in a surrounding agroforestry Buffer Zone. Here one finds small shrubs, medicinal plants, parasitic lianes for rope and furniture, and large trees that yield wood, edible leaves, and fruits.

Fifty percent of the biodiversity present in the primary forest can be found in an agroforestry Buffer Zone. In fact, says Hallé, it has only been in the past century that foresters recognized that the agroforestry Buffer Zone is human created and not a natural forest. Domesticated animals are grazed in the Buffer Zone, where the huts and villages are also located. The primary forest remains intact to provide new material during collecting expeditions.

Hallé says, "Agroforesters are true capitalists; their capital is biological, and it is constantly growing." Usually, they live off the interest, but when they are confronted with an emergency, they may harvest more than they usually take, sure in the knowledge that over time, the forest will grow back.

Hallé's description of agroforestry makes one wonder why it isn't being pushed everywhere as a sustainable alternative to massive clearing of tropical forests. Hallé's explanation is: "AF [agroforestry] is always local and small-scale. People are constantly coming out of the villages with baskets of fruits, vegetables, meat, and plant products for trade or sale, but that doesn't yield the large and quick profits that governments and multinational companies want."

Since all useful organisms are harvested from the Buffer Zone, the primary forest is protected as a priceless source of genetic material. Communities practicing agroforestry don't need outside help or expertise, because they depend on their own time-tested indigenous knowledge.

Hallé observes that practitioners of agroforestry are always women. Men may be recruited to cut down trees or lift heavy things, but women are in charge. He believes it reflects women's concerns with food and children's health. "Large-scale monoculturing seems to be more of a male impulse, while diverse, small-scale ventures seem more feminine," Hallé says.

Agroforestry exposes the insanity of destroying tropical forests for a one-time-only recovery of cash. Agroforestry rests on the fundamental capital of nature, which, if protected, can sustain communities and ecosystems indefinitely. But that flies in the face of the current suicidal path of global economics that glorifies human creativity and productivity above all.

Megadams

WE HUMAN BEINGS HAVE A REMARKABLE DESIRE AND ABILITY TO shape our surroundings for our convenience and comfort. And because we have the inventive and technological capacity to do it, any failure to exploit a natural "resource" is deemed a waste. Immense skyscrapers, bridges, and dams, drained swamps, and deep-sea oil rigs are a testimony to our engineering skills. But too often we fail to consider a project's impact beyond its immediate locale or payoff, and we end up paying heavy ecological costs.

In the debates about the future of dams at Alcan's Kemano Completion Project in British Columbia, Alberta's Oldman River, Saskatchewan's Rafferty and Alameda Rivers, and Phase II of Québec's James Bay Hydroelectric Project at Great Whale, critics often cite disastrous examples like Egypt's Aswan Dam on the Nile and Brazil's Balbina Dam in the Amazon rain forest. Let's look at northern British Columbia.

British Columbia's great river systems generate enough energy to supply the province's domestic needs and to allow it to export the excess. In the early 1960s, B.C. Hydro targeted the Peace River as an economic mother lode, and in 1967, the W.A.C. Bennett Dam (named after the longtime premier of British Columbia) was completed with great fanfare. Henceforth, the unpredictable water flow of the Peace could be "controlled" and regulated, subordinated to human needs. Few, if any, wondered about the dam's effects on a unique ecosystem 1,200 kilometers (745 miles) away in Alberta's Wood Buffalo National Park, famed for having the only known nesting sites of the near-extinct whooping cranes.

The heart of Wood Buffalo Park is the world's largest freshwater delta, formed by silt deposits from the Peace and Athabasca Rivers. The wetlands and meadows of the 5,000-square-kilometer (1,930-square-mile) area support millions of muskrats, of which up to 600,000 a year were trapped by the people of Fort Chipewayan. Over a million waterfowl and other migratory birds also exploit the tremendous productivity of the delta. And the sedge grasses provide nutritious feed for the bison for whom the park is named.

Dams change river ecology downstream, because the normal seasonal fluctuations in water levels are completely altered. In nature, rivers are low in winter and summer and flow heavily in spring. But electrical demands require peak release of water during the winter and less in the spring and summer. Before completion of the Bennett Dam, every five to eight years, ice would pile up in the Peace and cause a massive backup of water that would flood the delta.

Humans don't like floods, especially when they occur erratically, but for the delta ecosystem, those floods were vital. The flooding was completely dependable within the elastic cycle of time, and the plants and animals of the affected region were exquisitely adapted to and dependent on it. Floodwaters flushed out accumulated chemicals and drowned willows and other low shrubs that encroached when wet areas dried out. Silt left after the water receded fertilized the delta for the protein-rich sedge grasses that the bison herds depended on.

There hasn't been a flood in the delta for sixteen years, a triumph of our flood control. But a comparison of satellite photos made in 1976 and 1989 reveals a shocking change—40 percent of the productive sedge meadow habitat has been seriously altered and invaded by less palatable (for the bison) willows and low shrubs. Without the flooding to recharge the delta, 75 percent of the entire meadow area will be lost in the coming years.

Wood Buffalo National Park is a global heritage. One of the largest parks in the world, it is larger than Switzerland. It provides nesting grounds for whooping cranes and range for the largest free-roaming herd of bison on the planet. But it is under assault from within and without. Suggestions have been made to "depopulate" (meaning kill) the bison because they are infected with diseases that are said to threaten cattle. Magnificent stands of the tallest white spruce in Alberta have been logged for years in the park. And soon immense pulp mills in northern Alberta will spew toxic effluents into the air and water of the north.

The catastrophic impact of the Bennett Dam should teach us how little we know about the complex interconnections between ecosystems. Canada has some of the richest wilderness treasures in the world. But we must have greater humility about what we know and look far beyond immediate benefits and local effects or else we'll put it all in peril.

Update

Wood Buffalo National Park continues to be assaulted by human activity. In a study commissioned by the Canadian Park Service, the drying of the delta was reported to be continuing because of the prevention of floods by the Bennett Dam far upstream. At the rate of drying, the report predicted, the delta will disappear as a vibrant ecosystem in thirty years.

The bison herd was infected with bovine brucellosis and tuberculosis, causing cattle farmers in neighboring lands to fear the infection of their animals. They pressured the government to eliminate the threat from the bison, and a plan was devised to kill the entire free-roaming herd of more than three thousand animals and to replace them with uninfected animals from other parks. It was an astounding plan, considering the fact that the park is one of the largest in the world, encompassing 44,000 square kilometers (17,000 square miles). Public reaction to the proposal was swift and intense, forcing the government to back down and suggest an alternative scheme, namely, to round up the animals and cull only the infected bison. It's one thing to propose shooting all animals, but the scheme to capture, pen, and test thousands of wild creatures boggles the mind. To date, the plan has not been implemented.

Meanwhile, clear-cut logging within the national park continues, and pulp mills are polluting the air and the water in the park.

Global Warming

THE ENTIRE HISTORY OF VIRTUALLY ALL MODERN TECHNOLOGY, from combustion engines to rockets, nuclear power, and telecommunications, has occurred within the last 150 years. And the effect has reached the atmosphere itself.

Certain molecules in the atmosphere, including water and carbon dioxide, allow sunlight to pass through them but tend to reflect infrared or heat. Thus, like glass panes in a greenhouse, the gaseous molecules act to heat the Earth. The unprecedented increase and accumulation of human-created greenhouse gases (carbon dioxide, methane, CFCs) could cause forest die-off, desertification of farmland, and higher sea levels. And yet Canada, the United States, and other countries do little to reduce the output of the gases. The reasons for reluctance to reduce greenhouse gases are obvious—a serious effort will require a large investment, changes in lifestyle, and a fundamental shift away from the relentless priority of growth.

Some people question the seriousness of the dangers. The greenhouse effect is exaggerated, they argue. Fluctuations in global temperature have occurred in the past, and extinction is normal, since 99 percent of all species that ever lived are now extinct. These two points fail to consider the *rate* of change. The warming that ended the last Ice Age occurred at the rate of about a degree every millennium. The coming rate of change could be several degrees per *century*. And in the past, extinction rates may have been a species or two per year, while, according to an estimate in February 1993, we may be losing more than five species per *hour!*

A more serious criticism of people concerned about global warming is the lack of hard evidence to back up their fears. The fact is that naturally occurring greenhouse gases such as carbon dioxide and methane have been increasing in the upper atmosphere while new ones like CFCs have been created and released. Most climatologists believe that warming is already happening and will accelerate in the coming decades. But our ignorance about the factors that influence weather and climate is so great that it is impossible to make a realistic scientific prediction.

Because of those uncertainties, the Marshall Institute, a right-wing think tank in Washington, D.C., published a paper in 1989 that concluded that the temperature increase already observed over the past century has resulted from the sun's natural variations. With greater evaporation and cloud formation, the article suggested, the Earth would be shielded from sunlight and actually become cooler. The business magazine *Forbes* used the report to excoriate environmentalists for their alarmist exaggerations, while then–U.S. president George Bush was persuaded to oppose the imposition of any targets to limit emissions.

The Marshall Institute report had a widespread influence. An editorial in Canada's *Globe and Mail* (October 12, 1990) headlined "What We Don't Know About Global Warming" warned that action to avert global warming would have vast economic repercussions. Citing the Marshall Institute report, the editorial concluded, "In the absence of more solid information on the dimensions of the danger, the proposed insurance premiums seem out of proportion to apparent risk."

The interim report of the Canadian Parliament's all-party Standing Committee on Environment that was tabled in 1990 put the issue of global warming into its proper perspective. While acknowledging the uncertainties of climate prediction, the committee members "nonetheless accept the argument that the precautionary principle must apply in so vital a situation. By the time scientists have all the answers to these questions, global climate may have been driven by human society to the point where the answers are largely academic." The report goes on to warn of the reality of atmospheric change and sees "no validity in the argument that governments should delay acting until more detailed information on the likely effects of global climate change is gathered.... If the skeptics are correct and climate change is less of a problem than most scientists anticipate, the policies which we are proposing will still return many benefits, both environmental and economic."

The report indicated that some countries were already acting. "West Germany has adopted the target of reducing CO_2 emissions by 25 percent in 2005 from 1987 levels; Denmark and New Zealand will attempt to reduce CO_2 emissions by 20 percent in 2000 from 1990 levels ... the Committee concludes that the Toronto target—a 20 percent reduction in the 1988 level of CO_2 emissions by 2005—is the minimum that Canada should strive for as an interim goal."

Back in 1979 at the first World Climate Conference (WCC), experts wondered whether human-induced global warming could happen. But by 1988, enough was known to lead delegates at the Toronto conference to propose what has become a standard by which to ensure a country's seriousness in addressing climate change. A year later, at a meeting in Holland, the European Community favored a formal agreement to reduce emissions but was opposed by the United States, Britain, Japan, and the Soviet Union. The American delegation argued that the threat of global warming was still not certain, citing the report by the Marshall Institute. None of the authors of that report attended the Geneva Conference in 1995, where the Intergovernmental Panel on Climate Change (IPCC) was unequivocal in its conclusion about the scientific basis for global warming. Bert Bolin, chairman of the panel, said that his committee's findings "buried the Marshall Report for good." The IPCC documented the increase in atmospheric content of human-produced greenhouse gases since the Industrial Revolution. They concluded with certainty that this "will enhance the greenhouse effect, resulting on average in an additional warming of the Earth's surface."

The IPCC report stated "with confidence" that the relative effect of different gases can be calculated and that carbon dioxide (mainly from burning fossil fuels) has been responsible for over half the enhanced greenhouse effect and will likely remain so in the future. Since atmospheric levels of long-lived gases adjust slowly over time, "continued emissions ... at present rates would commit us to increased concentrations for centuries ahead. The longer emissions continue to increase at present-day rates, the greater reductions would have to be for concentrations to stabilise at a given level." In other words, it's far easier, cheaper, and faster to act *now* than to wait till later when even more gases will have been added to the upper atmosphere.

The IPCC suggests that CO_2, nitrous oxide, and CFCs have to be cut by "over 60 percent to stabilise their concentrations at today's levels." That won't be easy, but we haven't even begun to try. Several studies, including a government-commissioned Canadian document, have shown that there are huge environmental and *economic* benefits from cutting back on emissions. In Geneva, Thomas Johansson of Sweden reported that the United States could cut CO_2 emissions by 20 percent by the year 2000 and save a whopping U.S. $60 billion a year. He found that countries as diverse as Sweden and India could also reap enormous benefits from energy conservation. Around the world, countries are wasting energy and, in the process, adding greenhouse gases unnecessarily to the atmosphere.

Of all industrialized countries in the world, Canada has the poorest per capita record. With only 0.4 percent of the world's population, Canada generates 2 percent of all greenhouse gases. With near unanimity of expert opinion, Canadians have no excuses for delay.

If greenhouse gases aren't brought into equilibrium, the consequences will be catastrophic. For the web of life on Earth, even small shifts in average global temperatures will have huge repercussions. The atmosphere and air quality will change. Climate and weather will become even more unpredictable, and global patterns of snow, ice, and permafrost will be altered. Water supplies will be interrupted, and saltwater intrusion will contaminate drinking water on the coasts.

Terrestrial ecosystems will be affected even more than those in the oceans because water temperature changes more slowly. In a forest ecosystem, each species will respond differently. Some will flourish, whereas others will die out, so the collection of organisms will change. Agriculture will be thrown into chaos as growing areas, rains, and seasons are transformed unpredictably.

Both the World Health Organization (WHO) and the Australian Medical Research Council concluded that the effect of global warming on human health will be disastrous. For populations already at the edge of starvation, food shortages and increased costs will exacerbate their malnutrition and vulnerability to diseases. In the northern industrialized world, skin and eye diseases will increase, as will the incidence of tropical parasites and diseases.

The most predictable consequence of warming is the effect on the oceans. When water warms, it expands. When a mass of water as large as an ocean heats up even a bit, sea levels rise. As a result, ocean currents are changed, marine ecosystems altered, and plankton populations affected. Warmer oceans will increase the intensity of tides, storms, erosion, saltwater intrusion in aquifers, and corrosion of underground subways and pipes. A sea level rise of a few centimeters will greatly affect human societies. Coastal areas will experience storms of greater intensity and frequency. For people living in lowland deltas of Bangladesh, Egypt, and China, and on coral islands like the Maldives and Seychelles, the results will be disastrous. Permanent flooding will create millions of environmental refugees.

Fifteen of the twenty largest cities in the world are built next to oceans. The cost of countering sea level rise will be vast. Holland's famous storm barrages to prevent a recurrence of the killer flood of the 1950s cost over $8 billion, an amount that will be dwarfed by the efforts to protect cities and beaches around the world.

In New Orleans, we can glimpse the horrifying consequences of sea level rise. First let me remind you that marshes are vital parts of ecosystems that support many life-forms, including fish, mammals, and birds. But 50 percent of U.S. marshes have already been destroyed, and 40 percent of what remains is located in southern Louisiana around New Orleans. Today, Louisiana is losing 40 hectares (100 acres) of marsh and farmland a *day* to flooding.

Flying over what not long ago were rich marshes, we can see the expanse of green change to a series of ponds that become larger and larger. There is evidence everywhere of abandoned human effort—systems of levees, dams, and pumping stations that were used to keep back the rising water. Like modern-day King Canutes, people spent years raising levees, pumping longer and harder, and repairing breaches in dikes, only to retreat and abandon fields, houses, roads, and machinery. Their drowned dreams and hopes remain visible from the air.

Louisiana's loss of land began with the failure of people to pay attention to nature's rhythms as an essential part of the Mississippi delta. The delta was often flooded by hurricanes or river overflow. So in 1936, the Army Corps of Engineers, which is probably on a par with the World Bank for its ecologically destructive acts, completed an elaborate system of levees to stop the Mississippi from flooding. It may have been good for people's property, but it killed the delta. Over thousands of years, the Mississippi moved across the delta, flooding, carving new channels, tossing up mud banks, and replenishing the soil. Once channels were fixed, as oil, gas, and water were pumped from beneath the delta, it sank. Fishers in the area are enjoying a bumper yield of shrimp, crabs, and fish, but what they are doing is harvesting the organic material from the marshes. The rich soil that accumulated over thousands of years is now being broken up and flushed to sea, thereby fertilizing the waters and creating a short-term bonanza that will peter out.

Although sea level rise is occurring with astonishing speed in geological terms, it is invisible to most people. In spite of our capacity to plan ahead to avoid danger, we don't react to incremental change, only to major disasters like hurricanes or floods. Already sea level rise is a real problem in cities such as Venice, New York, Miami, San Francisco, Tokyo, and Osaka. We will be forced to pay enormous sums to counter the effects of this rise and protect cities in the industrial world, but developing countries will not have the money. And even worse, we don't appear to have learned from the lessons of New Orleans. The main conclusion from the 1995 climate conference in Geneva is that we know too little to make accurate predictions about the effect of rising temperature. But we can say with absolute certainty that cycles and regularities, like monsoon rains and seasons that people have relied on for tens of thousands of years, will no longer be dependable, while at the same time the topology of the Earth's ecosystems will change.

We are on the edge of a global catastrophe, and it's time politicians took the warnings of scientists to heart. We need vision and leadership—for the sake of our children.

Why We Must Act on Global Warming

"SO MUCH FOR GLOBAL WARMING," A FRIEND REMARKED WHILE DIScussing the frigid winter suffered in eastern Canada in 1994. With that dismissive comment, he had leaped from a single observation to a conclusion that is simply not warranted.

During the run of extremely hot summers of the late 1980s, concern about global warming reached a peak. But the fact that six of the hottest years on record occurred in the '80s did not constitute "proof" of global warming. That run of hot years could simply have been part of a normal pattern of fluctuations. Those hot years did not prove anything any more than one cold winter did. I reminded my friend that while the East was freezing, British Columbia was experiencing record high temperatures. Global warming is about global, not regional, temperatures.

It is going to take a lot more data collecting and hypothesizing to prove whether or not global warming really is a threat. So should we hold off doing anything, as many economists and businesspeople suggest, until the evidence is in? At a conference in Geneva in 1990, more than seven hundred atmospheric experts from all over the world agreed that we are putting unprecedented amounts of greenhouse gases into the upper atmosphere and that all signs indicate the world has warmed over the past century.

Human beings are adding more greenhouse gases into the upper atmosphere than can be removed annually. By trapping heat on Earth, greenhouse gases have kept the planet at a temperature in which life flourishes. More of these molecules will increase the effect.

The long-term consequences of this excess in greenhouse gases are hard to predict. As the Earth heats up, it is acknowledged, there will be greater evaporation from the ocean surfaces. This will lead to greater cloud cover and, as that spreads over the Earth, sunlight will be blocked. That, some conclude, will lead to a cooling rather than a heating of the Earth, and that is a reasonable hypothesis.

Others suggest that with rising temperatures, there will be greater turbulence in the air as different parts of the planet heat up at different rates. Convection currents could sweep upward and create tall columns of clouds rather than flat horizontal sheets. That would expose even more clear sky, so the Earth would heat up even faster. This too is a valid possibility.

The fact is we don't know what will happen. By tweaking parameters and factors in complex computer models of the atmosphere, we can get predictions ranging from an impending ice age to catastrophic heating. In view of the range of possibilities, many scientists suggest that the highest priority is to increase research funds so that evidence can be gathered to make better predictions. That suggestion is inadequate, however, if it is then assumed that we can carry on with business as usual until the data are all in.

Human numbers and technology have expanded to a point where we are changing the biophysical features of the Earth. With little from the past to guide our actions, we are playing a crap game with the only home we have.

We have to learn to live within the mechanisms that keep a balance among the components of complex ecosystems. In many areas of human activity, that means cutting back and hoping that the regenerative powers of nature will redress our damage. That's what we're doing with CFC damage to the ozone layer and overfishing off the Atlantic Coast. We have to reduce logging of old-growth forests and change agricultural practices that degrade soil and poison the land. And we have to decrease greenhouse gas output because the atmospheric changes, whatever they are, will have massive consequences and must be minimized.

The shocking fact is that studies in Canada, the United States, Australia, and Sweden all reach the same conclusion: a significant reduction of the output of carbon dioxide, one of the main greenhouse gases, will enhance health and the environment while saving billions of dollars. Unfortunately, the costs of reduction must be paid immediately, but the benefits accrue only years later, beyond the time frame of political vision. So there is little political incentive to do the best thing.

We have no choice but to act now to minimize the extremities of our uncertain future. But first, we've got to stop making assumptions about climate because of a change in temperature over a single week, month, or season.

Update

The fossil fuel industry continues to support a handful of naysaying scientists who speak out against the reality of human-induced climate change and therefore consider implementation of the Kyoto Protocol an expensive waste. But overwhelmingly, scientists throughout the world continue to call for serious action to reduce greenhouse gas production. The findings and recommendations of the Intergovernmental Panel on Climate Change, which includes more than 1,500 climatologists, have been strongly supported by the National Academy of Sciences of the United States, the Royal Society of Canada, and the Royal Society of London, among many other organizations representing the leading scientists of many countries. In December 2002, after a rancorous debate fueled by the opposition of Alberta's premier, Ralph Klein, the Canadian Manufacturers and Exporters Association, and the petrochemical industry, Prime Minister Jean Chrétien and the government ratified the Kyoto Protocol.

A study jointly supported by the World Wildlife Fund of Canada and the David Suzuki Foundation (*Habitats at Risk: Global Warming and Species Loss in Globally Significant Terrestrial Ecosystems*) showed that the warming that is already happening requires plants and animals to move north to remain within their optimal temperature range. Many plants simply cannot move fast enough, and animals often encounter open roads, fields, buildings, and fences that act as barriers to movement. The study concludes that most of the national parks in Canada will be significantly affected as their species composition undergoes radical change.

Another study, jointly produced by the Union of Concerned Scientists and the David Suzuki Foundation (*Confronting Climate Change in the Great*

Lakes Region), showed that the 60 million people living around the Great Lakes water basin will be strongly affected by global warming as well.

The warnings are many, yet there is also plenty of evidence to suggest that meeting the challenge of climate change by reducing greenhouse gas emissions is an opportunity. Reduced emissions will improve health, save money through energy conservation, and open new opportunities in alternative energy and in efficiency. *Power Shift: Cool Solutions to Global Warming* and *Kyoto and Beyond: The Low Emission Path to Innovation and Efficiency,* reports for the David Suzuki Foundation by energy expert Ralph Torrie, demonstrate that using technology already available (that means imminent technologies such as hydrogen fuel cells are not considered), Canada could reduce total greenhouse gas emissions by more than 50 percent in thirty years. Clearly, cutting-edge technologies that exploit energy like hydrogen will soon become available and make the job easier and faster. The initiatives needed to meet the challenge of climate change are not economic or technological; rather, it is necessary to overcome the mind-set that immediately assumes it can't be done or will be ruinously costly.

Ecological Footprints

THE STEEP RISE IN HUMAN POPULATION AND TECHNOLOGY HAS serious ecological repercussions that can't be alleviated simply by tinkering with social, economic, and political structures. We need to shift our priorities to find ways to live more lightly on the planet while ensuring that important human economic and social needs are not compromised.

Professor Bill Rees of the University of British Columbia and his associate, Mathis Wackernagel, confront us with stark facts that demand response. In *Our Ecological Footprint: Reducing Human Impact on the Earth,* they define our real needs: "Human life depends on nature's resource production, waste sinks and life-support services. Securing ecological stability is therefore a non-negotiable bottom line: nature's limited productivity is an ecological constraint within which humanity must live."

Rees uses a concept called the ecological footprint, which measures "a community's demand on the global carrying capacity and compares this with nature's available longterm carrying capacity. In other words, nature's productivity is compared with human demands.... In the long run, humanity cannot continue to consume more than nature produces. Human activities are bound to remain within the globe's ecological carrying capacity. To avoid the destruction of nature's productivity, humanity's ecological footprint must be reduced to the globe's carrying capacity."

Using this analysis, Wackernagel explores in more detail the basic human needs provided for by nature: "Energy is needed for heat and mobility, wood for housing and paper products, and we need quality food and clean water for healthy living ... green plants convert sunlight, carbon dioxide, nutrients and water into plant matter, and all the food chains which support animal life—including our own—are based on this plant matter. Nature also absorbs our waste products, and provides life support services such as climate stability and protection from ultraviolet radiation. Further, nature is a source of joy and inspiration."

A sustainable future depends on not using renewable resources more quickly than they can be restored and on not releasing more wastes than nature can absorb. "We know from the increasing loss of forests, soil erosion and contamination, fishery depletion, loss of species and the accumulation of greenhouse gases that our current overuse of nature is compromising our future well-being."

The Vancouver-based Task Force on Planning Healthy and Sustainable Communities developed methods to measure human consumption according to units of land needed to supply those services: "Appropriated Carrying Capacity or ecological footprint is the land that would be required on this planet to support our current lifestyle forever." Wackernagel says, "The ecological footprint of an average Canadian adds up to over 4.8 hectares [11.9 acres] or an area comparable to three city blocks." That includes 1.3 hectares (3.2 acres) for food, 1.0 (2.5 acres) for housing, 1.1 (2.7 acres) for transport, and 1.1 (2.7 acres) for consumer goods. Looked at another way, land use involves 2.9 hectares (7.2 acres) for energy, 1.1 (2.7 acres) for farmland, 0.6 (1.5 acres) for forest, 0.2 (0.5 acre) under pavement and buildings.

Wackernagel calculates the ecological footprints of different kinds of households: a single parent with child has annual home expenditures of CDN $16,000 and requires 3.1 hectares (7.7 acres); a student living alone needs CDN $10,000 and 3.9 hectares (9.6 acres); an average Canadian family (2.72 people), CDN $37,000 and 4.8 hectares (11.9 acres); a professional couple with no children, CDN $79,000 and 13.5 hectares (33.4 acres).

The ecologically productive land available to each person on Earth has decreased from 5 hectares (12 acres) in 1900 to 3.6 hectares (8.9 acres) in 1950 to 1.7 hectares or 4.2 acres (0.3 hectares/0.75 acres arable), in 1990. Assuming no further soil degradation and current population growth, it will decline to 0.9 hectares (2.2 acres) by 2030. Land appropriated by richer countries has increased from 1 hectare (2.5 acres) per person in 1900 to 2 (5 acres) in 1950 to 4 to 6 (10 to 15 acres) in 1990. Wackernagel's shocking conclusion is that "If everyone on Earth lived like the average Canadian, we'd need at least *three Earths* to provide all the material and energy essentials we currently use."

He goes on: "The Lower Fraser Valley, the area east of Vancouver, contains 1.7 million people, or 4.25 people per hectare. If the average Canadian needs 4.8 hectares [11.9 acres], then the Lower Fraser Valley needs an area 20 times what's actually available for food, forestry products and energy.... In other words, human settlements don't affect only the area where they're built."

Concentrating people in cities can, however, reduce the energy needed for transportation and housing. The challenge, says Wackernagel, is "to find a way to balance human consumption and nature's limited productivity in order to ensure that our communities are sustainable locally, regionally and globally. We don't have a choice about whether to do this, but we can choose how we do it. All of us are consumers of nature's productivity. We must work together to achieve a more sustainable way of living now in order to ensure that resources continue to be available not only for ourselves, but also for future generations."

Rees, Wackernagel, and their associates have provided us with a powerful tool to recognize the extent to which we are now exceeding the capacity of our surroundings to support us. If we expect that people in all countries have the right to aspire to our level of consumption, then clearly we are in for a major crisis. Sustainable living means coming into balance with the capacity of our surroundings to support us. The ecological footprint analysis provides a standard that should dictate our actions and politics for the future.

Update

Mathis Wackernagel continues to apply the Ecological Footprint (EF) analysis to compare demands in different countries. He works with Redefining Progress, an NGO in San Francisco. Today, it takes an average of 2.3 hectares (5.7 acres) to supply the average needs of every person now alive. But there are only 1.9 hectares (4.7 acres) of land available per individual.

Applying the EF analysis to the entire populations of different countries, it is possible to compare a nation's ecological impact in relation to the amount of land available in that nation. Thus, countries like Australia, Canada, and New Zealand, with very high EFs, can meet their needs from their land base because of relatively small populations and large land mass (see table). But even a large country like the United States, with the highest EF of all, has a deficit—that is, it cannot be supported by its own landbase.

Wealthy countries such as Holland, Switzerland, Japan, and Germany have lower EFs than the previously mentioned nations but require more land than they have within their borders. Surprisingly, countries with huge populations, such as India, China, and Bangladesh, remain close to meeting their own needs from their landbase because of their low consumption. As their economies continue to grow, however, that will certainly change.

Ecological Footprint Accounts

Country	Total Footprint (hectares per individual)	Biocapacity (hectares per individual)	Ecological Deficit
Argentina	3.0	1.9	–1.1
Australia	7.6	14.6	7.0
Austria	4.7	2.8	–1.9
Bangladesh	0.5	0.3	–0.2
Brazil	2.4	6.0	3.6
Canada	8.8	14.2	5.4
China	1.5	1.0	–0.5
Denmark	6.6	3.2	–3.4
Germany	4.7	1.7	–3.0
India	0.8	0.7	–0.1
Japan	4.8	0.7	–4.1
Netherlands	4.8	0.8	–4.0
New Zealand	8.7	23.0	14.3
Russia	4.5	1.6	–2.9
Switzerland	4.1	1.8	–2.3
United Kingdom	5.3	1.6	–3.7
United States	9.7	5.3	–4.4

In 2002, Wackernagel et al. looked at six categories of humanity's demand from land and sea: crops for food, feed, and fiber; raising animals; harvesting timber; fishing; infrastructure such as roads and housing; burning fossil fuels. They measured the amount of land it takes to supply those resources and absorb wastes. They found that in 1961, humanity's load on Earth was 70 percent of its productive capacity. In other words, what all people in the world required in those six categories in a year took 70 percent of the year to be replaced. By 1999, this had grown to 120 percent, and a mere three years later to 125 percent. It now takes one and a quarter years to replenish what we've used in a year, and this deficit has been going on, according to Wackernagel et al., for more than two decades. That means in order to meet our annual demands, we are exploiting our natural capital rather than living within the interest, or the regenerative capacity, of the planet. And just as a bank account may be drawn for more than the annual interest for years, eventually it will be exhausted and yield nothing. The EF provides a grim warning that we are living far beyond our ecological means.

This unsustainable demand has enormous implications for other species. On May 22, 2002, a United Nations Environment Program report warned that almost one-quarter of all mammalian species face extinction within thirty years. The primary causes are habitat destruction and introduction of alien species. One in eight bird species and more than five thousand plant species are also at risk. And this refers only to those species we know about. Biologists may have identified as few as 10 to 15 percent of all species on Earth, but the unknown species too are being lost.

A Boost for Biodiversity

FOR YEARS, ENVIRONMENTALISTS HAVE SOUNDED THE ALARM ABOUT the threat of species extinction. Whooping cranes, whales, spotted owls, and marbled murrelets have been at the center of debate about the fate of the biological makeup of all life on Earth.

The explosive rise in human population and consumption has accelerated the rate of species extinction. Our use of space for housing, food, materials, and energy has put enormous pressure on wild places and organisms. Complex ecosystems are transformed into biological deserts by our dams, farms, and housing developments. In imposing human power and priorities on our surroundings, we radically reduce biological diversity on the planet.

By our actions, we have shown little regard for the value of biological diversity, yet heterogeneity is a characteristic of populations of living organisms and is found at the level of DNA within every species. This genetic variability has been assumed to be responsible for the evolutionary persistence of a species by conferring resilience and flexibility. That's why there is concern for the fate of species such as bison and tigers, which have been reduced to a small number of survivors. With a reduced amount of genetic variation because there are so few individuals, they are more vulnerable to a new threat such as predation, disease, or environmental change.

Biologists have long theorized that biodiversity confers flexibility on species and ecosystems, enabling them to survive different assaults and bounce back from disasters. At the same time, it has been suggested that maintaining maximal species and ecosystem diversity isn't that urgent because many species are genetically similar. This idea implies that what really matters is representation of species within major groupings, because related species are ecologically redundant. In a time of explosive disputes over the fate of the Great Whale River and the Clayoquot forest, these opposing suggestions have serious repercussions.

In a paper entitled "Biodiversity and Stability in Grasslands," published in the prestigious scientific journal *Nature,* the ecologists David Tilman and John Downing accumulate strong evidence for a biological role of biodiversity. In a twelve-year study, they counted between one and twenty-six different species in 207 Prairie test plots, each measuring 3 meters by 3 meters (10 feet by 10 feet). For each plot, they determined the biological productivity by the weight of vegetation.

In 1987–88 the study region underwent a severe drought, and the researchers found that the productivity of the plots correlated strongly with the number of plant species. Vegetation shrank to about half of the predrought average in plots with the most species, and to about an eighth in the most species-poor plots.

Equally significant, the resilience of the plots after the drought was also correlated with the biodiversity. Thus, by 1992, the species-rich plots had returned to predrought levels of productivity, whereas the species-poor land still hadn't achieved those levels. The authors conclude, "This study implies that the preservation of biodiversity is essential for the maintenance of stable productivity in ecosystems."

Tilman says, "Biodiversity of an ecosystem has a major impact on its stability and functioning. This work leaves little doubt that biodiversity matters and that habitats with more species withstand stress better and recover faster.... The more species you have, the more likely some will be resistant to drought or other stress. We should preserve biodiversity because it's nature's insurance policy against catastrophes.... By sustaining biodiversity, we help sustain ourselves." Downing adds: "Instead of eliminating species from our forests, wetlands, roadsides, lake shores, powerlines, parks and lawns, we should be preserving them."

Based on his work, Downing warns: "Today humans are greatly reducing biodiversity, not just in the tropics but right here in North America.

More than 12 percent of plant species in the U.S. are already listed as rare or endangered, and in Canada, more than 1 percent of known species are threatened. Some countries may lose as many as 30 percent of their plant species over the next two decades. This loss of species has serious long-term implications for life on Earth. It makes the environment more susceptible to insects, disease, fire, drought, abnormally hot or cold weather, floods, acidification and other stresses."

Right now, as an example, rising temperature from global warming wreaks havoc in marine and terrestrial populations. Forests may have to move a thousand kilometers a century to stay within a tolerable temperature range. Forests can and do move, but normally at a tenth of that rate. Only biodiversity offers the hope of harboring the right combinations that can adapt to climate change. Unfortunately, most of that diversity resides in ecosystems that seem irresistible for human exploitation and development, which end up reducing variety.

And now humanity itself is being monocultured at an alarming rate. Media and advertising spread the perspectives and values of a single culture dominated by economics worldwide. Free trade, global competition, and uniform mass products of transnational corporations are blanketing the world with a notion of progress and development that has proved highly destructive of local communities and ecosystems. At this very moment, we need an infusion of other views, especially the attitudes and knowledge base of ancient, indigenous cultures.

But governments around the world seem determined to pull us into greater globalization and diminished diversity, even though current ideas of progress, wealth, and development are not enhancing the quality of life for the vast majority of people. It is in the diversity of local communities and local ecosystems that we will find the solutions for stability and a sustainable future. With concrete scientific evidence for the value of biodiversity, the question is, Does our society take science seriously?

Learning from Nature

"LOOK AT THAT INSECT," THE GRANDMOTHER SAID, POINTING AT THE beetle lying motionless on the sidewalk. "Oh, its battery must be dead!" responded the boy.

This story was related to me in Japan in 1994. Apocryphal or not, it was meant to illustrate how disconnected from nature modern people have become. To the boy, even an insect is merely an object manufactured by humans.

I thought of that anecdote while lying on a dock on the shores of Smoke Lake in Algonquin Park. The reason we have become so reckless with the ecological birthright of our children must be the illusion that we are in total command of our surroundings. Here in Algonquin, we have an opportunity to see that nature is complex beyond our understanding.

The brilliant panoply of stars overhead at night is a startling contrast to the limited display visible in any large city on a "clear" night. Gazing at the spectacular starscape above me, I wondered whether urban children responded to their first sight of an Algonquin night sky with awe or fear. I suspected a bit of both.

Our perceptions and values are formed by our experiences and surroundings. For example, Prairie people have told me of their great love of the big sky and the distant horizon. As a result, they feel claustrophobic in Vancouver, hemmed in by mountains that seem to push them into the ocean. For most people today, the human-created environment of cities shapes priorities and the way we see the world. Here in Algonquin, we have to recognize that there is a different rhythm, based on the seasons, the climate, the forest, and the animals, over which humans have no control at all.

"Tonight is 'wolf night' in Algonquin," someone said, to explain the people who had gathered to howl in the forest. The haunting reply of a wolf confirmed an interspecies communication and sent squeals of delight through the crowd. This is quite a change in attitude from that reflected in fairy tales where vicious wolves try to make a meal of Little Red Riding Hood or the three little pigs.

This summer I have been able to spend time relaxing on Quadra Island and rafting down the Babine River in British Columbia, traveling in the Arctic around Baffin Island and Greenland, and now holidaying in Algonquin, the jewel of Ontario's park system. Taken together, these places give an overwhelming sense of the awesome beauty of this part of the planet. But these islands of wilderness are decreasing in number and size under the relentless pressure of human beings. Too many of us want too much to be sustained indefinitely by the productive capacity of even the most distant ecosystems. Communities of organisms with whom we are discovering a spiritual kinship—grizzlies, walruses, wolves—are threatened by our need for "development" and "resources."

The degradation of natural systems has become inevitable because our economic system is fatally flawed by species chauvinism. Economists appraise everything in the world on the basis of perceived utility for human beings alone—if we can use it, it has a value; if we can't, it's worthless. That might work if implicit in such a system was an understanding that our survival and quality of life depend on what we extract from the Earth—air, water, soil, biodiversity. Unfortunately, that isn't how it works. Economists have traditionally defined those things—the very things that keep us alive—as externalities to the system they've invented. And therein lies the basis of our destructiveness.

And nowhere in such a system is there any acknowledgment of the importance and reality of spiritual values. The immense expanse we call Canada includes some of the most dazzling and biologically rich places on this planet. The chortle of a loon, which has inspired generations of Canadian artists and writers, brings me back to a reality too easily forgotten in cities. Loons and moose remind me how important it is to be able to share this Earth with other species. They inform us that we must temper our impulse to exploit and profit, because there are forces and phenomena that lie beyond our scientific understanding.

Away from Ontario's wilderness gem at Algonquin, we traveled to northern British Columbia. Rafting the Babine River along a 160-kilometer (100-mile) arc above Smithers, we are transported back in time. For four days, the only hint of the existence of other people is the whisper of a far-off jet. Away from the familiar urban setting, we live by a different rhythm and rules and see through different eyes. There is much to think about and time to do it.

"Daddy, is it safe to drink the water?" my daughter asks, reminding me how much our world has changed since camping in my childhood, when we drank from rivers and lakes without a worry. Her whoop of delight when I answer yes reassures me that the simple joy of being in a pristine environment persists.

Hundreds of eagles precede us in a steady relay. Drifting by rock strata pointing straight up, we can only speculate at the forces that caused these geological upheavals. And everywhere we wonder at trees that have somehow found a roothold on rock cliffs or whole forests growing out of thin topsoil coating solid stone.

From where we start just below Babine Lake to our finish at the Kispiox River, the water drops almost 500 meters (1,600 feet), creating both spectacular scenery and a navigational challenge. On the way, we meet the early arrivals of the sockeye salmon run. They leap past one of the hundreds of obstacles on the way to spawning channels almost a thousand kilometers (600 miles) from the sea. And all along the route, fewer and fewer evade the deadly string of predators depending on them.

We sight twelve grizzly bears, perhaps 10 percent of the population in this watershed, and watch three of them fishing in a narrow canyon. Wild and free, large animals like grizzlies need a lot of room to move. As powerful as they are, they are extremely wary of humans and almost always vanish with a shout. But as the relentless pressure of human activity encroaches on their homes, bears can't just move away or squeeze in tighter. I wonder how much room we will leave for these magnificent animals, for whom this is ancestral land.

At each campsite, the children are respectful but unafraid. They know the bears will give us a wide berth if we are careful and make enough noise to let them know we're here. River guides tell stories of urban dwellers for whom this kind of wilderness is terrifying because it is incomprehensible and unpredictable. Yet it is exhilarating and somehow reassuring to discover forces of nature still intact that are beyond our comprehension and domination.

For four days, the forest around us is intact, untouched by road or logging, but at the junction with the Skeena River we sight the first patches of clear-cuts. A massive fire in 1978 on Cutoff Mountain gave forest companies an excuse to "salvage" the burned trees by clear-cutting. Those companies often argue that fires and outbreaks of insects and disease are the natural analogues of clear-cuts, but after such logging there's nothing left to salvage. The unlogged parts of the Cutoff burn have much more vegetation and regrowth. Standing or fallen dead trees continue to hold the thin soil and nourish a host of organisms, including small trees.

Along the Skeena, fresh clearings far exceeding the legal limit of 50 hectares (124 acres) appear on both sides of the river. Forest companies boast that for every tree cut, two or three more are planted, as if a few seedlings crudely stuck in the ground are an equal exchange for trees hundreds of years old. A mature tree bears hundreds of cones, each producing hundreds of seeds. Of tens of thousands of seeds, perhaps a few hundred will actually sprout and begin to grow, and only a few dozen may reach small treehood. Of them, a handful may reach full maturity. Every merchantable tree, then, is a survivor of tens of thousands of mini-experiments, not two or three seedlings.

Those calling for the preservation of areas like the Babine River basin are called "greedy" because only a few hundred people may experience it annually. But when such an area is developed, an exquisitely balanced and complex ecosystem thousands of years in the making vanishes in a geological blink of an eye.

Human beings have needs that are more than just physical and social; we are spiritual animals. I believe we have a built-in need to experience wilderness and nature, a craving that can be fulfilled partially by knowing that there are such cathedrals to nourish the soul. In trashing wilderness areas that only time and nature create, we diminish ourselves with the loss of an integral component of our spiritual makeup.

When we began the trip at the head of the Babine River, we put our rafts in at a weir built by the federal government. Sockeye salmon completing their long journey from the sea up the Skeena River and then the Babine are stopped by that weir so that scientists can determine the number and makeup of the run and regulate the number of fish in the spawning beds.

Driven by the powerful urges of their genes, the fish gather restlessly behind the barrier until they find a series of slots through which they swim only to be trapped in metal cages. At regular intervals, summer students open a horizontal slit so that the fish can slither by and be counted and classed as mature or "jacks" (immature males).

Impounded in the metal cage, the fish frantically leap into the air, repeatedly smashing themselves against the metal sides. It was a horrifying reminder of a giant dam on the Paraná River in South America. Many of the fish in the Paraná migrate up and down the river, foraging for food or spawning beds. So a fish "elevator" was installed to carry the fish from one side of the dam to the other. Fish attempting to escape the elevator made the same sickening clank of flesh on metal as the desperate animals on the Babine. The elevator is a cruel and ridiculous technological solution to fulfill the natural instincts of fish. And here in Canada, in the name of science, experts were treating fish in a similarly unnecessary and destructive way.

Tourist rafting on the Babine has increased the human traffic considerably. This summer (1994), two or three hundred people will probably make the trip. Since there are very few sites along the rugged banks that can comfortably accommodate tents and rafts, there is increasing pressure on these sites. There is also concern that though wild, animals could become habituated to humans by rooting through any refuse or even digging up latrines.

Our rafting guides went to impressive lengths to minimize our impact. There wasn't a sign of human detritus such as packaging, bottle tops, or cans on the entire trip. Every bit of food that was brought in was consumed, carried back out unused, or burned and the ashes spread on the river. Only small branches of driftwood were collected and used for firewood so that they could be burned completely and the ashes dispersed.

Most surprising was the collection of all human feces so it could be transported away from the river. Of course we did have an effect on the ecosystem, but every effort was made to minimize it.

It was a privilege to enter territory where the wild creatures—eagles, salmon, and bears—so clearly belonged. Without even being aware of it, we showed our respect by our own behavior. When we encountered grizzly bears fishing, we waited and watched quietly until they had finished and left before we continued on our way.

At the Skeena River, we encountered our first evidence of clear-cut logging on both sides of the river. Along the riverbanks, a fringe of trees was left all the way up to the tops of the banks. But the sky was visible through the highest trees and the narrow strip of trees became an insult, an inadequate illusion of a real forest. There was a striking contrast between the riverbanks along the Babine and those on the Skeena. Slides where rock and dirt had come away and dropped toward the river carrying trees with them were not at all noticeable on the Babine—in fact, I didn't see any. But once we hit the logged area, slide after slide was obvious along the banks, many of them recent.

It may be true that the bottoms of natural slides are places where some of the largest trees in a wild forest can be found. When thin soil piles up at a slide's bottom, it can be an exceptionally fertile area for future trees. But along a river, the soil slides right off the hills and into the water, while the gashes remaining are inhospitable challenges for any future trees. So even where trees are left for cosmetic purposes, clear-cut logging nearby has visible effects.

In the wild, the interconnectedness of living systems becomes obvious. The salmon and steelhead runs up the Skeena and Babine are famous and important for sport, commercial, and Native food fisheries. But they are also intimately connected to biophysical and meteorological rhythms that are disrupted by large-scale clear-cutting. And the populations of bear, eagles, and microorganisms are intertwined with the fate of salmon. An experience of the wild uplifts the human spirit and also makes us rethink some of our assumptions and beliefs.

Near the End of Life

MANY PLANTS AND ANIMALS ARE SHOWING SIGNS OF PLANETARY DIStress. The ones that disappear or become rare are the frontline victims. But the warnings also appear in the fish covered with tumors, the bird chicks born with abnormalities, and our children's asthma and immune diseases.

I thought a lot about cancer and death a couple of years ago, after our family doctor detected signs of a tumor in my father. Dad had already survived an episode of cancer of the tongue and secondary tumors in his lymph glands. Thanks to excellent medical treatment, he was considered cured and had enjoyed eight high-quality years. But this time it was in his liver.

Approaching eighty-three, my father remained a solid rock for me. My severest critic and also my biggest fan and defender, he was a patient teacher for my children and the source of a lot of skills and knowledge I never acquired. As I knew he would, Dad took the biopsy results philosophically. "We all have to die. I've had a rich life and I have no regrets," he assured me.

Like Dad, each of us has to come to terms with our own mortality. He was just a lot further along his life, but I knew he would fight for what more he could get and he would relish every bit of it.

Our anticipation of death is the terrible burden. Yet painful as it is, the passing away of our elders is a natural and necessary part of life. In contrast, the death of a child or someone in the prime of life shocks us as an affront to the expected order. But today we are jeopardizing the sources of life itself—the air, water, and soil—by contaminating them with massive quantities of novel, human-created toxic chemicals.

However much we attempt to take responsibility for our health and bodies by diet and exercise, we have no choice when it comes to what is in our environment. Each glass of tapwater in Toronto or New York, for example, is estimated to contain several hundred thousand molecules of dioxin as well as dozens of other toxic chemicals. Each day our lungs filter thousands of liters of the same air into which industry and our vehicles discharge their effluents. In the name of efficiency and profit, the food we eat is laced with a variety of chemicals whose long-term persistence, accumulation, and physiological effects are not known.

No one knows the consequences of all of this, yet we are assured by politicians and experts that air, water, and food are safe. "Acceptable" levels of novel chemicals in the environment are arrived at with the best of intentions but are based on so many extrapolations, assumptions, and value judgments that they are hard to take seriously.

How do we judge, for example, whether water is safe to drink? The amount of a toxic chemical in water is divided by the maximum acceptable concentration. If the ratio is 1.0 or less, the water is presumed potable. But most water sources today contain dozens of toxic chemicals. Suppose it is found for a sample that the ratio is 0.8 (and therefore deemed acceptable) for PCBS, 0.7 for mercury, 0.28 for dioxin, and so on. Are we supposed to be assured that the water is drinkable? Shouldn't the ratios be added or even multiplied? They aren't. So what is the cumulative effect of all of these compounds? No one knows.

Crude levels of safety or hazard for a toxic chemical are established by time-consuming and expensive tests. But when it comes to assaying two or more different compounds simultaneously, testing breaks down. There are simply too many combinations and permutations of varying concentrations. Different compounds may also interact synergistically to produce a novel or much greater effect than expected on the basis of the properties of each chemical alone. Thus, we cannot predict the consequences of the sum total of the spectrum of chemicals present in our environment.

We are performing a massive experiment with the planet and ourselves. The chances are high that the most sensitive segments of our population will be rapidly growing fetuses, infants and children, and the vulnerable elderly whose defense mechanisms have been weakened by age.

Cancer is primarily a disease of old age, but even in my father's case it raises the question of whether there was an environmental cause that could have been prevented. My father had already had fourscore and three years. Will our children be so lucky?

Update

Dad lived for two and a half high-quality years after his brush with liver cancer. He finally died on May 8, 1994. He was lucid and free of pain right up to the last few hours of his life. I moved in with him in the last weeks and the rest of the family arrived to be with him at home. Before his death, he helped me write his obituary:

> Carr Kaoru Suzuki died peacefully on May 8th. He was eighty-five. His ashes will be spread on the winds of Quadra Island. He found great strength in the Japanese tradition of nature-worship. Shortly before he died, he said: "I will return to nature where I came from. I will be part of the fish, the trees, the birds—that's my reincarnation. I have had a rich and full life and have no regrets. I will live on in your memories of me and through my grandchildren.

Learning to Slow Down

THIS MONTH I REACH MY SIXTY-SEVENTH BIRTHDAY. YIKES! I WAS wondering where the first three months of 2003 went, but I should be asking what happened to my life. I vividly remember those turbulent months after I reached puberty and lost about half of my IQ. I still haven't gained it all back. I seemed to become a walking gonad thinking only about one thing, and back in the 1950s, few of us could do any more than think about it. Testosterone has a lot to account for.

As a teenager, weeks could seem an eternity, and life stretched on without end. I felt invincible, and retirement or the importance of a well-rounded life that was fully lived was the furthest thing from my mind. As a young man, I discovered a passion for science and threw myself into research as if there were no tomorrow. As a university professor, genetics was my life; it consumed me and provided my greatest highs and lows. Author Theodore Roszak once described this ability of scientists "to be carried away with an idea" as a double-edged quality, enchanting for the enthusiasm but frightening for its narrow focus.

Looking back over my life, I realize that it was my enthusiasm and willingness to throw everything into the moment that attracted students to the lab. They also loved the communal aspect of total engagement. We would work till the early-morning hours, excitedly arguing over ideas and results, generating reams of data, and dreaming up ever more elaborate experiments. Today I don't torment myself with recriminations for what I did. But there were costs—a broken marriage; relationships with lovers, students, and children that suffered from inattention; a narrow life. But I recognize that many of my shortcomings—that inward focus on my own passions, the excitement of the moment, the working as if there is no tomorrow—are also features of society as a whole that have created the current ecological crisis.

The European immigrants who created the United States began with a vast territory filled with natural resources. The qualities of those early settlers—rugged individualism, a search for new frontiers to conquer, and pride of nation—are deeply embedded in the American psyche. The enormous economic success of the U.S. has served as a model for emulation around the world and pulled ambitious immigrants from every quarter. Colloquial expressions reveal the attitudes that have come to dominate that society: "full speed ahead," "sink or swim," "the sky's the limit," "there's plenty more where that came from," and "that's the price of progress."

My children gave me the greatest gift of my life—grandchildren. Suddenly life didn't stretch ahead of me forever. I have reached a later stage of my life and now it is finite and short. I am no longer haunted by death but by the legacy bequeathed to my grandchildren and their grandchildren. I remember interviewing loggers in a MacMillan Bloedel cutblock on Vancouver Island for *The Nature of Things.* After being hollered at for a while, I retorted, "Environmentalists are not against logging or loggers. We just want to be sure your children and grandchildren will continue to log forests as rich as the ones you're cutting now." Immediately, one of the loggers cut in, "No f—ing way I want my kids to be loggers. There won't be any trees by then." In his concern about his next paycheck and meeting household expenses, that logger ignored what he obviously knew, that current logging practices are not sustainable.

So here in the last part of my life, I can only hope that from an elder's perspective, I can offer a bit of clichéd advice. Slow down and smell the roses. Recognize that we live in a world where everything is connected to everything else and so whatever we do has repercussions. There *is* a tomorrow and what we do now will influence what tomorrow we arrive at. We owe it to future generations to think about them before leaping ahead.

ECONOMICS AND POLITICS

POLITICIANS AND BUSINESSPEOPLE, EGGED ON BY THE MEDIA, consider the economy to be the "bottom line." If we don't have a strong, growing economy, they say, we can't afford all the services we depend on, such as health care, education, social safety nets, and even environmental protection. Thus, it is perceived that the economy is our highest political and industrial priority because it has become the source of everything that matters in our lives. Nature is considered subordinate to the economy, to be protected if we have the finances to do so. Politicians contort policies and commitments to keep that economy growing. They also believe that a global economy is essential and eagerly enter into international markets.

By placing the economy above the natural world in importance, the global economy is destructive of local ecosystems and local communities. We have to examine some of the most deeply held and cherished assumptions about conventional economics to come to grips with its destructiveness. And politicians must recognize that leadership and vision are vitally necessary at this time.

The Hubris of Global Economics

WHAT IS POVERTY? IN URBAN CANADA IN 1994, THE LOW-INCOME cutoff line for a family of four was CDN $31,071. North Americans living on social assistance often manage to have a TV set, a telephone, a refrigerator, even a car, commodities that are beyond the dreams of most people here in Papua New Guinea, where I am visiting as a guest of a local environmental group. Laura Martin, minister of finance and planning in East Sepik Province, told me that "the official average PNG [Papua New Guinea] income is about $500 [Canadian] but it's really more like $300." A poor Canadian would be wealthy here.

Today Papua New Guineans appear destitute. Yet not long ago, in this unique place, poverty was virtually unknown. The University of Papua New Guinea social scientist Nick Faraclas writes:

> Imagine a society where there is no hunger, homelessness or unemployment, and where in times of need, individuals can rest assured that their community will make available to them every resource at its disposal. Imagine a society where decision makers rule only when the need arises, and then only by consultation, consensus and the consent of the community. Imagine a society where women have control over their means of production and reproduction, where housework is minimal and childcare is available 24 hours a day on demand. Imagine a society where there is little or no crime and where community conflicts are settled by sophisticated

> resolution procedures based on compensation to aggrieved parties for damages, with no recourse to concepts of guilt or punishment. Imagine a society ... in which the mere fact that a person exists is cause for celebration and a deep sense of responsibility to maintain and share that existence.

Such a place is not fiction, says Faraclas:

> When the first colonisers came to the island of New Guinea, they did not find one society that exactly fit the above description. Instead, they found over one thousand distinct language groups and many more distinct societies, the majority of which approximated closely the above description, but each in its own particular way. These were not perfect societies. They had many problems. But after some one hundred years of "Northern development" ... nearly all of the real developmental gains achieved over the past 40,000 years by the indigenous peoples of the island have been seriously eroded, while almost all of the original problems have gotten worse and have been added to a rapidly growing list of new imported problems.

When Columbus "discovered" the New World, the people he encountered were better off physically, materially, and spiritually than his crew members' families back in Europe. Poverty is a state of mind, and people may discover they are poor only when others tell them or when they see immense wealth flaunted by others.

The anthropologist Helena Norberg-Hodge told me that when she arrived in Ladakh twenty years ago, her first impression was of overwhelming hardship and poverty. Yet when she asked a boy to show her the poorest family in a village, he looked puzzled and finally replied, "There are no poor people here."

Over time, Norberg-Hodge came to appreciate the incredible wealth of kinship and community and the richness of culture and tradition in Ladakh. Urged by industrialized countries, however, the government decided that the country needed "development." So a road was built across the Himalayas to open the tiny nation to the outside world and bring in money, goods, and tourists. Norberg-Hodge watched stable communities break down as young people left for the allure of city life and material goods. She saw the same village boy, now a man, living in the capital city and begging for money from tourists because "we are poor."

World leaders who met in Bretton Woods, New Hampshire, in 1944 envisioned a new economic order, and to help achieve it they created the International Monetary Fund and World Bank. Now we can see that they were astonishingly successful. A mere fifty-one years later, the planet has been saturated with a single notion of development and economic progress. Governments big and small, from socialist to capitalist, military dictatorships to kingdoms and democracies, repeat like a mantra the faith that the global market and economy will improve living standards and bring wealth and opportunity.

Yet global economics is ultimately destructive because it is fatally flawed: it externalizes the natural capital and services that keep us alive while glorifying human inventiveness as if it allows us to escape finite limits and manage our biophysical surroundings; it assumes that endless growth is possible and necessary and represents progress; it does not value long-term social and ecological sustainability; it rejects caring, cooperation, and sharing as irrational while promoting selfishness; and it cannot incorporate the reality of spiritual needs.

It is breathtaking hubris to force this single, monolithic concept as salvation into every part of the world.

The Ecosystem as Capital

OUR FAILURE TO RECOGNIZE THE IMPOSSIBILITY OF MAINTAINING exponential growth in consumption and the economy is creating a global crisis of catastrophic proportions.

In the summer of 1988, Toronto was host to both the Economic Summit and a conference on the Changing Atmosphere. Prime Minister Mulroney attended and made pronouncements at both. At the Economic Summit, he reiterated the importance of maintaining economic growth, and at the atmosphere meeting, he urged action to avoid further destruction of the atmosphere. And he acted as if there was no connection between them! We continue to allow the same old concerns for growth, jobs, and profit to determine the political and social priorities of government. Leaders of all major parties are hemmed in by pressure groups, ignorance, and their personal value systems and seem to have neither the time nor the inclination to rise above the immediate exigencies of political survival and long-held political truths. Political "vision" seldom extends beyond the interval between elections, yet the environmental crisis must be seen on a longer scale.

It is *economics* that now preoccupies the media and politics. Maintaining growth in the economy by carving out a presence in a global economic community has become the raison d'être of almost every government in the world. Implicit in any economic system are arbitrary and irrational human values about work, profit, and goals. They alone render economic systems opaque to prediction and hence beyond "management." But today the magnitude of the global trade and monetary system has taken us into new dimensions of complexity for which there is no historical precedent—we simply have no idea where we are heading.

Global economics must be exposed for what it is—a complete perversion. To begin with, economics is a chauvinistic invention, a human creation based on a definition of value solely by the criterion of utility to our species. As long as we can see a use for something and hence can realize a profit from it, it has economic worth. Yet it is the ecosystem that is the fundamental "capital" on which all life depends. Financial leaders manipulate the monetary system for immediate profit with little regard for environmental or human consequences. The current climate of laissez-faire economics in which the marketplace and private sector are being released from government constraints only ensures greater environmental depredation. We are only one species out of perhaps thirty million, and however much we may think we are outside nature and control it, as biological beings we remain as dependent on clean air, water, and soil as all other organisms. Economics has no ecological foundation because it dismisses air, water, soil, and biodiversity as limitless "externalities" shared globally.

Our preoccupation with profit deflects us from taking effective action on such issues as the greenhouse effect, acid rain, salmon depletion, forest destruction, or contamination of fresh water in an ecologically sensible way. Countries like Canada and Australia, whose natural resources are the envy of most people on the globe, squander their natural treasures in a rush to maximize profit. Even though we have barely begun to understand the scope of diversity of life on Earth and the complexity of atmosphere, oceans, and soil, the drive for profit subsumes any concern over the long-term implications of our ignorance. These days when entire ecosystems are destroyed by clear-cut logging, hydroelectric dams, farming, or urban sprawl, we offer money to the dispossessed as if cash can compensate for things that are unique and irreplaceable.

Modern economics is perverted by an addiction to military spending. Each year, over a trillion (U.S.) dollars are spent worldwide for defense, including the manufacture, sales, and use of machines of death. (That's about $20 million every second.) The weapons trade consumes scarce resources but generates enormous profits for major industries and nations.

But it is the diversion of scientific creativity by the military away from socially and environmentally useful areas that is the greatest perversion.

Military research and development now consumes more than 71 percent of all U.S. research and development, whereas only 9 percent is spent on health research. According to R.T. Sivard in *World Military and Social Expenditures,* 1987–88, "the U.S. government devotes well over twice as much research money to weapons as to all other research needs combined—including energy, health, education and food—and the Soviet pattern is believed to be similar."

Global economics is perverted because it impoverishes much of the Third World by seducing its people with the blandishments of technological "progress." High-tech weaponry, disposable goods, highly mechanized agriculture, and substitutes for mother's milk have had devastating social and economic effects on less developed countries. To pay, Third World countries mortgage their future by selling off irreplaceable capital—their natural resources. Brazil, for example, has teetered on the brink of economic collapse for years. To keep up with a massive international debt, the country is destroying a unique world treasure, its rain forest. It is the richest ecosystem on Earth, home to much of the biological diversity of the planet and affects the atmosphere, soil, water cycles, and climate in ways we don't understand. It is criminal to destroy those forests merely to service the interest on international debt, for when they are gone, Brazil will still be mired in debt.

If our candidates for office make claims to genuine vision and concern for the future, they must be made to answer profound questions about the interrelationship between the environment and economics. For the sake of our children, we cannot afford to go on with business as usual.

Update

Research into ecosystems and conservation can provide tools and techniques with which to measure and counter ecological degradation. But Harvard's E.O. Wilson once told me that more money is spent in two weeks in bars in New York City than is spent annually to do basic descriptive taxonomy around the world. And the expenditure that dwarfs all commitments to scientific research, medicine, and education is for the military. In the twenty-first century, with all of our sophisticated technology and scientific discovery, we divert a grotesque amount of money to machines of destruction.

The special Congressional bill for the war in Iraq in early 2003 alone was U.S. $74 billion! Around the world, there are more than 22 million military personnel, 70,000 in Canada alone. Globally, the total military expenditures exceed U.S. $1 trillion. In 2001, military expenditures in Canada were U.S. $7.9 billion (representing 1.1 percent of the GDP); in Sweden, U.S. $4.4 billion (2.1 percent of GDP); and in the United States, U.S. $276.7 billion (3.2 percent of GDP). Wouldn't the world be a more secure place in the long run if our military budget was switched with our budget for environmental commitments?

Ecologists and Economists Unite!

THE WORDS *ECOLOGY* AND *ECONOMICS* DERIVE FROM THE SAME Greek word, *oikos,* meaning "household" or "home." So ecology (*logos* meaning "study") is the study of home, and economics (*nomics* meaning "management") is home management. These two fields should be companion disciplines, and yet with few exceptions there is little communication between them.

Even though the fundamental capital that all countries depend on is the natural world, modern economics makes no ecological sense. When a businessperson evaluates a forest, for example, that ecosystem is transformed into "board feet" or "cubic meters" that can then be plugged into the proper equations. Other factors—costs of surveying and road and bridge building, labor, reforestation, market demand, and profit—are weighed before deciding whether a forest is worth logging. But considerations of the worth of not touching the forest are dismissed as "externalities" to economic calculations.

For most of human existence, we could get away with thoughtless extraction of "resources" from the environment because of the abundance and diversity within the natural world. Our numbers were small and technology simple and powered by human and animal muscle power. (That was still enough to build the pyramids and the Great Wall of China and to transform a number of forests into deserts.)

The invention of machines and exploitation of cheap, plentiful fossil fuels created a sudden and massive increase in technological muscle power that has had enormous ecological repercussions. Today, our species alone has the power to affect the other thirty million on the planet. Almost overnight, we can destroy entire ecosystems. But conditioned by the long-standing resilience of nature, we've continued to act as if it is virtually limitless, and this behavior is reflected in economic systems.

The planet is being ravaged for economic returns. But any farsighted economist must recognize that there are "services" performed by nature itself that have to be factored into the economic equations. So let's start by remembering that we are animals. As biological beings, we must have clean air, water, and food for our sustenance and health. The biological world around us has assured us of that. In the past, pollution by our fires, left-overs, and body wastes were recycled by other organisms. Today, the sheer magnitude, variety, and novelty of our technological excreta preclude that.

The great forests of the world have served to modulate the water cycles of the planet, absorbing rain and transpiring it into the air or releasing it into the ground. Thus, groundwater, erosion, flooding, landslides, and weather are directly affected by forests. Forests also absorb carbon dioxide while releasing oxygen, thereby conditioning the air we breathe and the upper atmosphere that affects climate. Old-growth forests maintain a high degree of biodiversity on which long-term ecological stability depends. All these "services" continue to be performed as long as the trees are left standing, yet none of them is cost-accounted before a forest is cut down.

There are other benefits of natural systems for humans that are seldom assessed economically. The most obvious is the enormous human capacity to discover and then exploit other species. Many of our most powerful medicines still are biologically based. The vast pharmacopeia of traditional medicines and yet-to-be discovered plants in tropical rain forests promise far greater returns than the much-ballyhooed biotechnology.

Throughout history, people have used perhaps 7,000 kinds of plants for food, yet there are at least 75,000 edible plants, many superior to ones we currently use. Only about 150 have been grown commercially, but human nutrition today is based on only 20 or so major crops. (Of these, 3 grass species, rice, corn, and wheat, are the most important.) There are also very real esthetic, spiritual, and philosophical values for nature that are never reckoned in any economic model. If the forest industry can be compensated when forests are preserved for parks, why shouldn't society be compensated for the potential lost when trees are logged?

The late American economist Julian Simon complained that ecological critiques perpetuate a myth of scarcity and dwindling resources. Simon clearly states the absurd faith held by most economists: "There is no reason why human resourcefulness and enterprise cannot forever continue to respond to impending shortages and existing problems with new expedients that, after an adjustment period, leave us better off than before the problem arose."

Lester Brown, president of the Worldwatch Institute, countered: "The lack of ecological awareness has contributed to some of the shortcomings in economic analysis and policy formulations." Brown points to fisheries, forests, grasslands, and croplands as key areas for the global economy:

> The condition of the economy and these biological systems cannot be separated. As the global economy expands, pressures on the earth's biological systems are mounting. In large areas of the world, human claims on these systems are reaching an unsustainable level, a point where their productivity is being impaired. When that happens, fisheries collapse, forests disappear, grasslands are converted into barren wastelands, and croplands deteriorate along with quality of air, water, and other life-support resources.

Economists cannot afford any longer to ignore their companion discipline of ecology.

Economic Fallacy

ECONOMICS IS A DOMINANT PART OF OUR LIVES, BUT THE TRUE ECOlogical cost of our economics is never factored in. In fact, some indicators of economic health mask the real costs.

Our economy is made possible by the fact that we, as biological beings, exist on Earth's productivity. Yet we are told that we need continued economic growth to afford a clean environment. So we rip up the Earth's productive capacity in order to keep on growing, even though this conflicts with the most fundamental rule in economics—you don't spend all of your capital if you want to avoid bankruptcy.

Economists have provided us with various ways to assess the "health" of the economy. One of them, the Gross National Product (GNP), is the total market value of all goods and services in society created in a year. The GNP is a sacred measure of annual economic growth and positively encourages environmental degradation. Athough a standing old-growth forest, a wild caribou herd, an unused pure aquifer, a deep-sea vent, and an undammed watershed have immeasurable ecological values and perform countless "services" in the total planetary biosphere, they do not register in the GNP. Only when people find a way to exploit them for financial returns does the GNP go up.

The GNP is also devoid of assessments of the social and environmental costs associated with the increase in goods and services. Suppose a major fire at a chemical or nuclear plant or pollution from a pulp mill spreads toxic compounds over a vast area and many people become very sick. More nurses, doctors, hospitals, janitors, medicines, and so on will be needed—so the GNP goes up. If people begin to die as a result of that exposure to toxic substances, there's greater demand for undertakers, caskets, flowers, air travel for mourners, grave diggers, and lawyers—the GNP rises further! As Ralph Nader has said, "Every time there's a car accident, the GNP goes up." The GNP is so preposterous that it went up in 1989 because of the *Exxon Valdez* oil spill, which was the greatest marine disaster in American history.

The GNP does not even register the quality and quantity of clean air, water, soil, and biological diversity. And what about the things that don't result in the exchange of money? The very glue that keeps the social fabric of communities and families intact does not involve money and therefore is invisible to the GNP. The person who opts to be a full-time parent does not register economically, whereas paid baby-sitters, nannies, and day-care workers do. All of the volunteer services performed at many levels of society—care for the elderly, the disabled, or the mentally handicapped—do not appear in the GNP. One of my associates belongs to the Lions Club and spends weeks every year preparing for a road race for quadriplegics. He enjoys it immensely, and his actions provide severely handicapped people with some excitement and fun. The value to the community of this kind of volunteer work is beyond price yet does not contribute to the GNP. The pre-eminence of the economy and GNP tears at these hidden social services in developing countries and impels them to pursue cash to service international debt and purchase products of industrial countries.

The role of the GNP in disrupting the social and environmental underpinnings of industrialized nations is illustrated in a story in the magazine *Adbusters* (volume 1, number 3):

> Joe and Mary own a small farm. They are self-reliant, growing as much of their food as possible, and providing for most of their own needs. Their two children chip in and the family has a rich home life. Their family contributes to the health of their community and the nation … but they are not good for the nation's business because they consume so little.

> Joe and Mary can't make ends meet, so Joe finds a job in the city. He borrows $13,000 to buy a Toyota and drives 50 miles to work every day. The $13,000 and his yearly gas bill are added to the nation's Gross National Product (GNP).
>
> Then Mary divorces Joe because she can't handle his bad city moods anymore. The $11,000 lawyer's fee for dividing up the farm and assets is added to the nation's GNP. The people who buy the farm develop it into townhouses at $200,000 a pop. This results in a spectacular jump in the GNP.
>
> A year later Joe and Mary accidentally meet in a pub and decide to give it another go. They give up their city apartments, sell one of their cars and renovate a barn in the back of Mary's father's farm. They live frugally, watch their pennies and grow together as a family again. Guess what? The nation's GNP registers a fall and the economists tell us we are worse off.

I am not an economist, but you don't have to be one to know something is wrong and has to be changed.

A Progress Indicator That's Real

THE GDP, NOW ROUTINELY MEASURED INSTEAD OF THE GNP, IS AN economic register of the total value of goods and services bought and sold. It has become the key indicator of the health of our national economy and, for some, a measure of progress.

But the GDP only adds and never subtracts; that is, no distinction is made between destructive and productive activities. An industry that makes a profit while polluting a stream adds to the GDP. People poisoned by the polluted water and hospitalized will need the services of doctors, nurses, and lawyers, as well as goods like medicine and flowers, all adding to the GDP. When the polluter is made to clean up the problem, often with government subsidies, that too is added to the GDP!

In this value system, the destruction of an old-growth forest by clear-cutting, the flooding of a valley by a megadam, and the elimination of fish by drift-netting are all treated as income generation rather than depletion of an asset. Voluntary services, friendship, sharing, cooperating, and caring, all acts that create a sense of community, do not contribute to the GDP. Nor does the GDP factor in income distribution, unemployment, or the debt incurred to foreign countries. The GDP has no value for wise decisions made to ensure the long-term stability of communities or ecosystems.

Robert Kennedy, Jr., has said, "The GNP counts air pollution and cigarette advertising and ambulances to clear our highways of carnage.... Yet [it] does not allow for the health of our children, the quality of their education, or the joy of their play.... It measures neither our wit nor our courage; neither our wisdom nor our learning; neither our compassion nor our devotion to our country; it measures everything, in short, except that which makes life worthwhile."

An organization called Redefining Progress has issued a call for new measures of progress, which states: "Much of what we now consider economic growth, as measured by GDP, is really the fixing of blunders from the past and the borrowing of resources from the future." The organization proposes a new value called the Genuine Progress Indicator (GPI). The new indicator is an attempt to redress the deficits of the GDP by measuring the general well-being and sustainability of a nation. The techniques for calculating each component of the GPI are contained in Redefining Progress's book, *The Genuine Progress Indicator: Summary of Data and Methodology.*

In contrast to the GDP, the GPI contains adjustments for:

- Resource depletion—The loss of wetlands, farmland, and minerals (including oil) is a cost.
- Income distribution—"The GPI rises when the poor receive a larger percentage of national income, and falls when their share decreases."
- Housework and nonmarket transactions—Child care, cooking, cleaning, and home repairs are GPI gains.
- Changes in leisure time—"The GPI treats an increase in leisure as a benefit and decrease in leisure as a cost."
- Unemployment and underemployment—These are costs.
- Pollution—This is a GPI cost.
- Long-term environmental damage—It is treated as a cost.
- Life span of consumer durables and infrastructure—This is calculated on the length of service rendered by an item rather than the money spent on its purchase.
- Defensive expenditures—The price of maintaining a level of service without increasing it is a cost in the GPI. "The amount of money spent on commuting, the medical and material costs of automobile accidents, and the money that households are forced to spend on personal pollution control devices such as water filters are examples."

- Sustainable investments—"If a nation allows its capital stock to decline, or if it finances its investments out of borrowed capital rather than savings, it is living beyond its means."

And what does Redefining Progress find in comparing the GDP and GPI? The GDP per capita in the U.S. rose steadily from a per-person average of U.S. $7,865 in 1950 to U.S. $16,414 in 1992. The GPI reveals a different picture, rising from U.S. $5,663 in 1950 to a peak of U.S. $7,441 in 1969, then declining to U.S. $4,426 in 1992. By striving for continued economic growth, we have catastrophically lessened our quality of life, as indicated by the GPI.

Even a former president of the World Bank, Barber Conable, is aware that GDP values "represent an income that cannot be sustained." He says, "Current calculations ignore the degradation of the natural resource base and view the sales of nonrenewable resources entirely as income. A better way must be found to measure the prosperity and progress of mankind." I couldn't agree more.

Endless Growth— An Impossible Dream

TRY THIS. ASK A POLITICIAN WHY HE OR SHE IS RUNNING FOR OFFICE. Ask what the objectives of her or his party are. Chances are the responder will recite the goal of maintaining a healthy economy, maximizing growth, and carving out a share of the global marketplace. Our society and government are driven by these economic imperatives. Indeed, the way we measure success and progress is by economic indicators.

Yet this preoccupation with economic growth has blinded us to a far more important truth—economics itself is an invention that makes no ecological sense. By setting our priorities around economics and following the advice of economists, we hurtle along a catastrophically destructive path.

We pay little attention to the incredible complexity and interdependence of components of ecosystems of which we, as biological beings, remain a part. And we have almost no idea what the consequences of a vast human population with its enormous demands are on the biosphere. In economics, the bottom line is profit.

Judith Maxwell, when president of the Economic Council of Canada, admitted in an interview that economists simply haven't paid attention to ecological factors. Economists consider the environment to be essentially limitless, endlessly self-renewing, and free. As the eminent Stanford ecologist Paul Ehrlich remarked, "Economists are one of the last groups of professionals on earth who still believe in perpetual motion machines." The facts are that we live in a finite world; that human beings are now the most numerous large mammal and our numbers are increasing explosively; and that our technological inventions permit extraction of resources at an accelerating rate. To economists, these facts simply offer greater opportunity to expand markets and increase profit.

In economics, the role performed by components of natural communities is of no importance. For example, although a standing forest provides numerous ecological "services" such as inhibition of erosion, landslides, fires, and floods while cleansing the air, modulating climate and weather, supporting wildlife, and maintaining genetic diversity (I haven't even considered the spiritual value of forests for human beings), to economists, these are "externalities," irrelevant to their calculations. As the head of a multinational forest company once told me, "A tree only has value once it's cut down." That's a classic economic perspective. Economists live in a land of make-believe. They aim at steady growth in consumption, material goods, wealth, and profit as if it can be sustained indefinitely. And they have faith that human ingenuity will open up new frontiers for steady expansion while providing endless solutions to problems we create.

Economic growth has become an end in itself, a mindless goal that is sought by every country in the world and is the very measure of progress. Economists and politicians of the industrialized countries claim strong growth is necessary so that they can afford to help the poorer nations. Yet the rich countries, which have only 20 percent of the planet's population, consume 80 percent of its resources. Our consumption is the primary cause of global ecological destruction. Since the 1950s, the world has undergone an unprecedented growth in industry and the economy. Yet in the period of high growth between 1968 and 1978, the wealthiest 20 percent of the world gained 67 times as much in income as the bottom half. If the developing countries continued to increase their per capita wealth at the 1973–75 rate of 4.6 percent per year, they would need 100 years to reach the current levels in industrialized nations. And if the rich countries grew at "only" 3 percent in the same time, they would consume and produce 16 times as much as they do now and total world output would have grown twentyfold.

The Brundtland Commission report *Our Common Future* clearly described the global ecocrisis and the grotesque disparities between rich and poor countries. Sadly, the Brundtland report then went on to accept the preposterous notion that continued economic growth in industrialized countries is critical for the Third World and projected yet another round of economic growth that could lead to a fivefold increase in global wealth. Can you imagine having and using five times as much as you have today?

Human intellect cannot endlessly find new resources or create alternatives, because the Earth and its potential are finite. Disproportionate, unsustainable growth in the rich countries is destroying the planet. And to what end is such growth necessary? Not to achieve goals such as social justice, equality, sense of community, meaningful work, or a clean environment. Instead, we think growth is necessary because it has become our definition of progress. Surely we need to define more worthy values, goals, and priorities.

Brundtland's report contained a phrase, "sustainable development," that has been embraced by business and governments because it promises the best of all worlds, continued economic "development" (which most interpret to mean "growth") and protection of the environment. But as long as development is synonymous with economic growth, "sustainable development" is a cruel oxymoron.

BARCLAYS HALF-YEAR PROFITS PLUNGE 78% (*The Guardian,* August 7, 1992).

HUDSON'S BAY PROFIT DROPS 29% and TD PROFIT DROPS AS LOAN LOSSES TAKE TOLL (*Globe and Mail,* August 28, 1992).

SILICON VALLEY HAS MIDLIFE CRISIS AS ITS GROWTH BEGINS TO TAPER OFF (*Seattle Post-Intelligence,* September 29,1992).

Implicit in the stories beneath the headlines is the assumption that the reported decline in growth is bad news. But the first three stories did not describe net losses suffered by the companies, only a decline in amount of profit. Apparently, there is no limit to the levels of profit that can be aspired to, and progress is measured by whether profit levels continue to increase. The last story bemoaned the fact that after three decades of explosive growth in high-tech companies in Silicon Valley near Stanford University, the number of new businesses being started there is slowing. Even though each new company adds pressure on land and space in the area, growth forever seems to be the aim.

These stories illustrate a madness that now afflicts us—the belief that steady growth not only is a measure of society's progress but is absolutely essential. Few question whether this goal can be sustained, because money is regarded as something real and as an end in itself. Thus we honor people just because they possess vast amounts of money. Kaiapo Indians of the Amazon call money "dirty paper," which is a far more accurate description of currency, namely, paper that used to stand for real things of value.

In 1992, the rhetoric of the U.S. presidential race and the debate about the Canadian constitutional referendum focused on economics. Success or failure of a government is evaluated on its economic record on job creation, GNP, or national debt. Inflation and unemployment affect all of us, but we must examine the vast ecological repercussions of our unquestioning assumptions about both the necessity and the possibility of maintaining steady growth.

On a trip to England in October 1992, I found the British media obsessed with the instability of the pound, the integration of Europe's monetary system, GATT, and the consequences of North American free trade. During the disastrous slide in the pound that month, the British government pumped billions from its reserves to stop the decline. Electrons flowing through wires to computers, fax machines, and telephones flooded stock exchanges with enormous amounts of money, thereby making and losing vast amounts for investors. Yet none of this activity had anything to do with creating or adding value to the world. Money for its own sake has become a medium of value that is no longer anchored to reality or even put to use for humankind. Money can be amplified endlessly just by buying and selling money.

But everything that we use in our homes, offices, and playgrounds, whether energy, plastics, metal, or food, came from the Earth. We are Earth Beings who have an absolute requirement for air, water, soil, and other living things for our health and survival. Thus, the biophysical elements of the planet are the fundamental capital that make our society and lives possible and should be the foundation for all our spiritual and material values, including economics. And simple economics dictates the protection of capital while living on interest for long-term survival and stability.

But economics does not begin with a recognition that there are limits to growth, nor does it admit the fundamental importance of air, water, soil, and biodiversity. Long-term environmental costs of development or non-human values are not factored into our economies. Today, economic growth has become synonymous with progress, and since every society aspires to progress, there is never an end to our desire for more.

The natural world on which we depend is far too complex to be understood let alone managed. If the great rain forests of British Columbia "add fiber" (that is, increase in size) at 2 to 3 percent a year, then in principle, by logging less than 2 to 3 percent of the wood volume annually, the forests and logging should be sustainable forever. But the system of money that we have invented is no longer rooted in biophysical reality and grows far faster than nature's regenerative rates. Forest companies don't operate on five-hundred-year logging plans, and cutting only 2 to 3 percent of the trees is not good economics. To maximize short-term profit, it makes economic "sense" to clear-cut a forest and invest the profit for a far greater return. Furthermore, since money is liquid, it flows freely across political boundaries to other forests, and when trees are gone, the money can be put into fish. When the fish disappear, the money can be converted to computers or biotechnology. Thus, trees, fish, lakes, and rivers are "liquidated" in the name of immediate economic benefits.

We need a completely different accounting and value system that can bring us back into balance with the realities of the Earth. And the first place to start is by recognizing that steady, mindless growth, as is also the case with cancer, can be deadly.

Three Economists

ECONOMISTS BELIEVE THAT THE GLOBAL ECONOMY IS LIKE AN immense machine made up of pipes and cogs and levers controlling input and output from various sectors. If all its components are regulated, people believe that our economic engine will hum like a well-tuned car. It is a dangerous assumption because economics externalizes air, water, soil, and other life-forms, eventually relegating them to minor roles in human life.

Human activity has already significantly increased global temperature. So what do economists have to say about it? Here is the response of Yale University professor William D. Nordhaus in "Greenhouse Economics" in *The Economist* (July 7, 1990). He acknowledges that "scientific monitoring has firmly established the build-up of the main greenhouse gases." But as an economist, Nordhaus thinks planetary climate is irrelevant to urban life: "Cities are increasingly becoming climate-proofed by technological changes like air-conditioning and shopping malls." The economic importance of industrialization apparently renders Americans exempt from the physical world: "Greenhouse warming would have little effect on America's national output. About three percent of American GNP originates in climate-sensitive sectors such as farming and forestry. Another 10 percent comes from sectors only modestly sensitive—energy, water systems, property, and construction. Far the largest share, 87 percent comes from sectors, including most services, that are negligibly affected by climate change."

Even though our very lives depend on nutrition that comes from agriculture that is very climate-sensitive, to Nordhaus, food can be written off: "For the bulk of the economy, however—manufacturing, mining, utilities, finance, trade, and most services—climate change over the next few decades is likely to have less effect than the economic reunification of Germany this summer." His reasoning is similar to concluding that since the stomach is a minor organ compared with the brain, we can do without it.

But what about the urgent warnings about the consequences of increasing release of greenhouse gases? Here's Nordhaus's response: "All these prognostications are judgements based on immense uncertainties. They could be dead wrong. This uncertainty must affect mankind's choice of responses to the threat of global warning."

Since Nordhaus, like most economists, places the economy above all else when it comes to matters environmental, he finds the *economic* costs of meeting the greenhouse effect head-on too great. Nordhaus calculates the bill for a 60 percent reduction in greenhouse-gas emissions at over U.S. $300 billion annually. With a faith in the capacity of human ingenuity to solve our problems and improve our lives, Nordhaus's choice to counter global warming is technology: "The option of climate engineering has been completely neglected. Possibilities include shooting particulate matter into the stratosphere to cool the Earth, altering land-use patterns to change the globe's reflectivity, and cultivating carbon-eating organisms in the oceans." He then dismisses the fact that "such measures would raise profound legal, ethical, and environmental issues" because "they would also probably be far more cost-effective than shutting down the world's power plants."

There you have it, folks, the world seen through the bizarre lenses of an economist.

For another illustration of the economic perspective, let us turn to Julian Simon, an economics professor at the University of Maryland. He was a highly influential member of U.S. president Ronald Reagan's Council of Economic Advisers. In Simon's book *The Ultimate Resource,* he revels in the infinite capacity of human imagination and inventiveness. As an example of the basis for his optimism, he said in an interview in 1992 that coal and copper were nothing but black or shiny rocks, respectively, until people thought of ways to exploit them. "All these resources come out of our minds. This is where the resources are rather than in the ground or in the air."

But if you ask Simon about scarcity of finite resources, he retorts, "What scarcity?" Pointing out that life expectancy over the past two centuries has risen from thirty to over seventy years in industrialized nations, he says, "All resources have been getting more available rather than more scarce ... the prices of all natural resources have been falling rather than increasing. And our air and our water in rich countries have been getting cleaner rather than dirtier over the past decades. So all the facts show us things have been getting better with respect to human welfare rather than worse."

Taking oil as an example, Simon says there are vast amounts to be recovered simply by being more efficient, and "if we want it, we can make more oil" using energy from the sun and carbon dioxide so that "even oil, even energy is not finite in any meaningful sense."

Since World War II, there has been an unprecedented increase in standard of living, utilization of resources, and human population. Noting their parallel rise, Simon concludes, "There's no known reason why this cannot go on forever." Furthermore, he says, "There is no population problem," because more people means more human minds, so life will only get better and better.

He believes there are no limits to growth as long as there are creative people with challenges to motivate them. If we ever do exhaust finite resources, Simon says, people will come up with alternatives, or we can travel to other parts of the universe to find what we need. He concludes: "On our own planet, we have all the resources, and even all the knowledge that we would ever need to sustain more and more people forever. We don't need to go to the other planets or to space for that. But what we need [them] for is to provide us challenges that call out the best of our human spirit. To allow us to explore and to find excitement of the mind and of the soul. That's what we need them for."

Simon's is a make-believe world built on the fantasy of infallible and inexorable improvement of the human condition through sheer intellect. He fails to acknowledge human ignorance of the interconnectedness of all parts of the fabric of life or the long-term repercussions of what is an aberrant blip of rising numbers and consumption.

Julian Simon is a mainstream economist. Critics of his kind of economics are often discounted because they lack credentials. But Herman Daly, a professor of economics at Louisiana State University and senior economist with the World Bank, cannot be dismissed so easily. Among his books are *Steady State Economics* and *For the Common Good,* and in an interview for *The Nature of Things,* he was highly critical of his field while offering ideas of where we should be heading.

According to Daly, the trouble can be seen in the first pages of any economics textbook: "You get what's called the diagram of the circular flow of exchange value from firms to households, households to firms. Exchange value goes round and round. Nothing comes into that system from outside, nothing exits to the outside." Thus, the economy is regarded as a self-contained system, isolated from matter or energy.

To Daly, that makes no sense and would be comparable to a biologist suggesting that animals don't require air, water, or food. That's the stuff of a perpetual motion machine, which the laws of physics tell us is impossible. Yet that has been the economic perspective: "We thought of the economy as the total system and Nature as just a sector in the economy. So there's nothing to constrain the expansion of the total system and if Nature happens to get scarce, well, we'll just substitute some other sector."

Even though everything human beings use comes from the Earth, economists render the environment an abstraction, an "externality," or a mere component of economic systems. Daly says we must "start from an entirely different pre-analytic vision, namely the economy as an open sub-system of a larger, finite, materially closed, and non-growing ecosystem." Once seen as a subsystem, "then the first question is how big is the economy, the sub-system, relative to the total system. How big can it be? How big should it be?"

To Daly, economic wisdom would begin by recognizing what is physically impossible: "We really do face real constraints.... It's impossible to create matter and energy. It's impossible to travel faster than the speed of light. It's impossible to generate living things from non-living things. And so on. So impossibility theorems are not just negative, pessimistic statements. They're important recognitions of the world."

Daly points out that by neglecting finite limits and our dependence on the planetary biosphere, economists are able to assume that growth can go on forever. Yet, he warns: "It's impossible for the entire 5.4 billion people in the world to live at a level of resource consumption per capita equal to that of the U.S., Canada, western Europe ... we're already straining life support capacity, the regenerative and absorptive capacities of the ecosystem, beyond their sustainable limits."

We have to rethink our current notions of economic health and recovery, says Daly: "We've moved from an era of economic growth into an era that we might call anti-economic growth. That means that expansion of the physical scale of the human economy now increases environmental costs faster than it increases production benefits." We are constantly told that the Holy Grail of government and the private sector is economic growth. But if the ecological costs rise faster than benefits, growth is, in fact, making us poorer instead of richer.

In discussions about "sustainable development," it becomes clear that most people use development and growth interchangeably. Daly makes a major distinction: "Growth means the increase in size by assimilation or accretion of materials. So when something grows, it gets physically bigger. Development, by contrast, is the realization of potential, the evolution towards a greater, fuller, or better state. So it's quantitative expansion versus qualitative improvement.... Perhaps we can develop forever. But we certainly cannot grow forever."

Daly points out, "As human beings expand and take over the niches of more and more other species, [and as] we preempt a larger and larger total of life space for our uses, then there's going to be less left over for everything else.... We are living by an ideology of death. We're pushing into the capacity of the biosphere to support life." And it is being done, says Daly, to maintain life of "extravagant luxury."

Yet there are vast disparities in wealth both between and within nations. If we do not choose to write off the poorest, then, Daly believes, "the wealthy of the world are going to have to reduce their levels of material consumption, seek satisfaction in dimensions of life that do not require so much material throughput ... lower our own load on the ecosystem." That's so the poor can reach a measure of sufficiency. It means "movement from the top and from the bottom towards a kind of mean which is sustainable."

Current economic policy, says Daly, based on stimulating growth so that wealth will trickle down to the poor, doesn't work. But it would be political suicide to impose draconian measures and manage a shrinking economy. So instead, "whoever says technology will enable us to grow forever gets a good hearing because it's welcome news" and lets us avoid the drastic measures needed.

The marketplace does have a role in the future, but not on its own, says Daly. The collective action of society and government must set the conditions, forcing the marketplace to operate within fixed ecological limits, distributing income justly, and establishing economic sustainability. "Within those limits, the market can work very well to find efficient allocations."

Daly believes the challenge is to encourage local economies that make ecological sense, but, he warns, they are usually swept aside by transnational corporations intent on globalization of free trade and free capital mobility. This globalization results in:

> a weakening of national boundaries for economic purposes. National boundaries become permeable. Capital flows freely. Goods flow freely. People flow much more freely. And in the world in which factors of production, labor, and capital are highly mobile, then ... all of the virtues of comparative advantage are out the window.... The problem with globalization and free trade is that it greatly diminishes the power of the nation which is the fundamental unit of community.... It reduces national boundaries in strength and increases the relative power of transnational corporations. There is no world government to control transnational corporations. They're legally subject only to the laws of the various national governments under which they operate. They play off one against the other, thereby diminishing the strength of community, of nations relative to transnational capital.

To Daly, it makes much more sense "to go back the other way and have much more national capital, much more limited to the national domain under control of national governments." This confers far more stability and resilience than a single monolithic integrated, interdependent world system. Taking his cue from biological systems, he says, "Globalization, building just one tight integrated system, is very much akin to monoculture in agriculture. You're putting all your eggs in one basket instead of having a lot of biodiversity." That increases the risks and uncertainties and enhances the chances of failure.

To the suggestion that the global economy is inescapable, Daly retorts, "We don't have to compete in the global economy. We can put tariffs on goods. We can produce for local consumption. We can limit our international trade to whatever degree we want to." Globalization to Daly means competing at the lowest level—cheapest labor, child labor, environmental degradation, no social insurance, no medical service, et cetera. "There's this tremendous standard-lowering effect of unlimited world competition ... it's a really dumb idea ... we're going to compete away all of the virtues of the standard of living ... in a crazy drive to be competitive internationally."

To Daly, the economic and ecological crisis is ultimately spiritual because "the culture of consumption has been very dissatisfying ... materialism is only good up to a point. And then beyond that we need spiritual, cultural, intellectual pursuits." If we are going to bring economics into some kind of realistic integration with ecology, mainstream economists are going to have to reckon with the criticisms of Daly and begin to rethink with new insight and direction.

Economics and the Third World

ONE THING THAT WAS MADE VERY CLEAR DURING THE EARTH SUMMIT in Rio in 1992 was the inseparable link between economics, poverty, and environmental degradation. The human and ecological tragedies overwhelming Somalia, Haiti, Nepal, and Bangladesh are a canary's warning to the rest of the world, while the economic and industrial plans of India, Brazil, China, and the former Soviet Union and its allies have vast implications for the planet's ecosphere.

Foreign aid is not a frill or indulgence of the rich countries but a necessity to ensure a future for our children. Susan George is an economics analyst who specializes in the developing countries. In her books *How the Other Half Dies* (1976) and *A Fate Worse Than Debt* (1988), she makes a powerful case that the misery of the "South" is created by global economics. George believes the current faith in the global economy is a religion based on dogma that obliges "everyone to believe in this doctrine in order to be saved." The problem is that "We're letting the economy be outside the ecology and do whatever it likes. The economy has become the guiding principle ... powerful institutions like the World Bank and the International Monetary Fund are in a position to say if you don't believe our doctrine, you will not be saved. You will have no new loans. You'll have no opportunity to participate in the world system."

The reason economics is a disaster is that it isn't founded on the finite limits of a biophysical world. The economic institutions created by the rich countries are failures: "People are more miserable. More people are marginalized. More people are excluded. More people are going hungry. And they've even failed in financial terms. The debt is much greater than it was when they began imposing all these systems."

In George's opinion, "We are waging war on the Third World. The debt has become an instrument to keep these countries under control.... It is a political tool which has obliged all of the debtor countries to toe the line and to do the will of the northern creditors. It's much more efficient than colonialism."

Now consider George's devastating analysis of what the global economy has already done for the South:

> In 1982, the whole of what is still called the Third World, the southern hemisphere, owed $900 billion in debt. From 1982 until the end of 1991, that's 10 calendar years, these countries paid back $1.496 trillion in debt service. Their only reward for paying back that amount of debt was to find themselves owing $1.478 trillion, an increase of 64 percent over what they owed in 1982. In other words, you can't win.
>
> If you take the debt of subsaharan Africa over that same decade, you find an increase of 123 percent in spite of the fact that subsaharan Africa somehow, and at enormous human costs, scraped together nearly a billion dollars every month for its creditors. Subsaharan Africa, according to the OECD, paid back $950 million on average every month for 120 months between '82 and the end of '91. And their reward is to be 123 percent more in debt than they were 10 years ago.... The very poorest, too, paid back $300 million a month, every month. And they're 150 percent more in debt than they were. So you simply cannot grow out of your debt.
>
> [All figures in U.S. dollars]

That statistic echoed Brazil's lament that in the last three years of the 1980s, $50 billion was exported merely to service the interest on its debt.

How did we arrive at such a state? In the 1960s, the transnational corporations began pushing into the poor countries by offering a dazzling array of goodies under the rubric of "development." Consequently, George says, "every country in the world is being encouraged or, indeed, forced to integrate into the world economy instead of finding our major necessities of life at the local level or the regional or the national level. We're first going to the international economy and then coming down towards the local and the domestic."

This pattern represents a complete reversal in the way societies have operated in the past when the focus of attention was the local community. All people "got their major necessities from the local, domestic arena, then the slightly larger regional, then national. And finally, if necessary, they would go to the international level. You don't have enormous transport costs or ecological costs. You certainly don't have this integration of so many Third World countries on such inhuman and unremunerative terms that they have now."

George's thesis echoes that of World Bank senior economist Herman Daly, who believes that the thrust toward the global economy is exactly opposite to what we should all be doing. To Daly, transnationals undercut community values such as social justice, equal opportunity, sustainability, or happiness with their focus on obtaining short-term returns and maximizing profit. As Daly says, "The community is the level at which people actually know each other and in which they are able to make decisions and feel the consequences of those decisions."

Susan George believes we can begin to change this destructive bent by bringing "the major needs for not only our physical lives but our cultural lives much closer to home.... Think locally in order to act globally.... To save things everywhere, you've got to start by saving them somewhere." In the short run, that injunction may cost more but will bring long-term benefits to the community.

In spite of her devastating analysis of the crippling effect of debt on the South, George sees rays of hope: "People are reacting. People are not taking this structural adjustment lying down.... They are forming their own groups to combat this. NGO [nongovernmental organization] activity has never been so strong in the Third World. There are 1,600 different environmental groups in Brazil who came to the Rio conference.... The creativity of these groups in the South absolutely puts us to shame. They have a lot of inventiveness." George recommends that we in the rich countries support this creativity and the local priorities of Third World NGOS.

The fate of the poorest countries of the world has a direct effect on us through what George calls the Debt Boomerang. We are feeling the effect of the South's debt because "they're selling off their forests and that means that we are losing a climate stabilizer. When the trees go, that means that global warming is going to increase. It means that biodiversity is going to go." And globalization of the economy has meant that "in the United States, the job loss, directly due to the debt crisis, has been in the order of two million jobs at a very conservative estimate and in Europe it has been at least three-quarters of a million jobs."

Constant repetition of the mantra of growth, competitiveness, and globalization keeps us from dealing with the underlying issues that Daly and George describe.

Consumption as a Deliberate Goal

OUR USE OF RESOURCES, FROM OIL, GAS, WOOD, AND MINERALS TO air and water, has escalated dramatically in this century. Both individually and collectively, in cities and in nations, our consumptive demand has magnified our impact on the planet. But this rise in consumption has also been an integral part of the economy. It wasn't always this way. As early as 1907, the economist Simon Nelson Patten was severely criticized for his prescient warning of a change that was taking place: "The new morality does not consist in saving but in expanding consumption."

Consumption, a word that once meant to waste away under the effects of disease, now affects the planet as a central part of our economic system. What makes it insidious is that our own identities have become tied up in the need to have more. As Paul Wachtel says in *The Poverty of Affluence,* "Having more and newer things each year has become not just something we want but something we need. The idea of more, of ever increasing wealth, has become the center of our identity and our security, and we are caught up by it as the addict by his drugs."

According to Allen D. Kanner and Mary E. Gomes, writing in "The All-Consuming Self," "The purchase of a new product, especially a 'big ticket' item such as a car or computer, typically produces an immediate surge of pleasure and achievement, and often confers status and recognition upon the owner. Yet as the novelty wears off, the emptiness threatens to return. The standard consumer solution is to focus on the next promising purchase."

I am often told, "Well, it's human nature to want more. You can't buck that." It's true that when members of the Kaiapo tribe from the Amazon visited me in Vancouver, they saw a lot of things they wanted. If they hadn't found anything they wanted, it would have been a terrible indictment of a total emptiness of the way we live. But the Kaiapo lived deep in the Amazon forest in complete self-sufficiency. In contrast, our consumption is far beyond anything necessary for survival, and society's hyperconsumption is driven by the billions of dollars spent annually to make us want things. Bill Gates spent more than one billion dollars in advertising alone for Windows 95.

It is instructive to remember that the Great Depression of the 1930s came to an end because World War II provided a massive economic jolt. American industrial might was fanned to white heat to support the war effort, but as victory began to loom the business community worried about how to keep the economic boom going. The solution was consumption. Shortly after the war, the retailing analyst Victor Lebow declared:

> Our enormously productive economy ... demands that we make consumption our way of life, that we convert the buying and use of goods into rituals, that we seek our spiritual satisfaction, our ego satisfaction, in consumption.... We need things consumed, burned up, worn out, replaced, and discarded at an ever-increasing rate.

By 1953 the chairman of President Eisenhower's Council of Economic Advisers would state that the American economy's "ultimate purpose" was "to produce more consumer goods."

But there's a problem. When products are made to be durable, industry will eventually saturate the market. Solutions such as planned obsolescence and constantly expanding new markets to the Third World, elders, yuppies, children, specific ethnic groups, and so on have worked well to overcome this defect, at least temporarily. Coca-Cola president Donald R. Keough expressed a religious attitude to market opportunity: "When I think of Indonesia—a country on the equator with 180 million people, a median age of 18, and with a Moslem ban on alcohol—I feel I know what heaven looks like." The ultimate innovation to ensure an endless market has been disposability. Citing convenience or hygiene to justify products that are used once and thrown away, industry creates an endless market for those products.

Assigning a Value to Nature

EVERYTHING WE DEPEND ON FOR SURVIVAL, INCLUDING AIR AND water, soil, plastic, energy, glass, metal, and food, comes from the Earth and represents "natural capital." Yet we are led to believe it is a strong, growing economy that ensures that we receive these necessities. In the looking-glass world of economics, human capital is overvalued, whereas natural capital and the processes by which the living world maintains itself are undervalued or ignored altogether.

A schematic representation of the economy is filled with items and arrows indicating the intricate relationship between resource extraction, processing, manufacturing, retailing, and regulations, taxes, and incentives. The ozone layer, underground aquifers, topsoil, biodiversity, and fresh water are depicted as externalities: outside the realm of the economy. But that means the economic system is no longer grounded in the real world, since those so-called externalities are actually the life-support systems of the planet. Without them, there could be no life—and certainly no economy.

This grotesque distortion of reality by conventional economics has disastrous results, as E.F. Schumacher, an economist and author of *Small Is Beautiful*, says:

> The wider human habitat, far from being humanized and ennobled by man's activities, becomes standardized to dreariness or even degraded to ugliness. All this is being done because man-as-producer cannot afford "the luxury of not acting economically"

> and therefore cannot produce the very necessary "luxuries"—like health, beauty and permanence—which man-as-consumer desires more than anything else. It would cost too much; and the richer we become, the less we can "afford."

In the "household" of the living world, each system, each entity, has a part to play in the "economy" of the whole. A standing tree performs numerous ecological "services" for the Earth, yet none of these services has economic worth according to the way our system does its accounts. Economics recognizes the values in a forest such as its recreational and medicinal properties, but all value is still defined in terms of human use, not in terms of the entire living world.

Progressive economists like Herman Daly are attempting to include factors that were previously thought to be outside the boundaries of the economy—that is, to internalize factors that were once considered external. Now ecologists are attempting to document and evaluate those vast services rendered by nature.

If we try to replace or substitute a natural service with a human-created technology, we may get an estimate of its economic worth. Thus, we can compute the worth of water purification by a watershed by calculating the cost of doing the same thing with a purification plant. Some of nature's services can never be replaced, because we simply don't have the technological competence to even try. For example, David Pimentel calculates that on a sunny day in New York State, insects pollinate a trillion flowers, a feat no human technology can reproduce. Nevertheless, it is possible to make crude attempts at putting a dollar value on much of what nature does. When Robert Constanza and his associates did this, they came up with an annual economic value of about U.S. $30 trillion, an amount that is almost twice the collective annual GDPs of all the countries in the world! In our economic systems, these services aren't even part of the discussion, so it's no surprise that we are so ecologically destructive. Critics, of which I am one, of this economic assessment of ecological services warn that there are some things that are beyond economic worth, that might be considered sacred. In imposing an economic value on all parts of nature, there is a danger that economists will simply try to factor everything in, including air, water, and biodiversity, as if it can all be rationally calculated.

Nevertheless, by trying to estimate the economic worth of nature's services we quickly realize its enormous "value," a value that dwarfs our economies yet is currently unacknowledged. So we continue to push the limits of a finite planet. And still the process continues: the underlying assumptions and priorities still drive us on toward disaster. ❧

Toward More National Economies

LIVING IN A WAY THAT ENSURES A FUTURE FOR GENERATIONS YET TO come will require major changes in the way we organize ourselves and live. But at present, all our political and economic leaders are taking us down the same destructive path, toward maintaining growth in consumption, over-exploiting renewable natural resources, overseeing economic growth while employment declines, and supporting globalization. How can we bring the global economy that enmeshes all of us into alignment with the natural world on which it depends? Few economists even consider the question.

An exception is Herman Daly, who left his position as senior economist with the World Bank after six years. On returning to academia, Daly delivered a remarkable farewell speech full of important suggestions for the World Bank.

His first suggestion? "Stop counting the consumption of natural capital as income. Income is by definition the maximum amount that a society can consume this year and still be able to consume the same amount next year. That is, consumption this year, if it is to be called income, must leave intact the capacity to produce and consume the same amount next year. Thus, sustainability is built into the very definition of income."

Conventional economics forgets or ignores the fact that everything we depend on for survival comes from the Earth and represents "natural capital." Fiscal responsibility dictates that we live on interest and not touch the capital. But Daly points out that economists focus on the notion that wealth is created by human beings, so "productive capacity that must be maintained intact has traditionally been thought of as manmade capital only, excluding natural capital. We have habitually counted natural capital as a free good.... [I]n today's full world it is anti-economic."

Daly tells us that the error of considering natural-capital consumption as income is built into our system of national accounts, in the evaluation of projects that deplete natural capital and in the accounting of international balance of payments. He says the accounting "biases investment allocation toward projects that deplete natural capital, and away from more sustainable projects." Clearly this bias must be corrected if we are to have a sustainable future. Depletion of nonrenewable resources, or their exploitation beyond a sustainable yield, should be counted as user costs. So should the ability of air, water, or soil to act as a sink for the absorption of products of human activity, such as excess carbon dioxide.

Daly says, "In balance of payments accounting, the export of depleted natural capital, whether petroleum or timber cut beyond sustainable yield, is entered in the current account, and thus treated entirely as income. This is an accounting error. Some portion of those nonsustainable exports should be treated as the sale of a capital asset.... If this were properly done, some countries would see their apparent balance of trade surplus converted into a true deficit, one that is being financed by drawdown and transfer abroad of their stock of natural capital." Had this been done for the northern cod or Pacific Northwest old-growth forests, our economic and political decisions might have been very different.

His second recommendation is to "tax labor and income less and tax resource throughput more. [The matter and energy that go into a system and eventually come out are what goes through—the throughput.] In the past it has been customary for governments to subsidize resource throughput to stimulate growth. Thus energy, water, fertilizer and even deforestation are even now frequently subsidized." In most countries, labor and income are taxed, but in so doing we merely exacerbate the pressure toward higher levels of unemployment. Instead, Daly says, we should be moving our tax base *away* from labor and income.

Daly goes on: "Income tax structure should be maintained so as to keep progressivity in the overall tax structure by taxing very high incomes and subsidizing very low incomes. But the bulk of public revenue would be raised from taxes on throughput either at the depletion or pollution end." He points out that such changes must be instituted in the North first. The problem is that the World Bank has leverage to encourage environmentally sustainable development only in the countries it gives loans to, namely, the poor countries of the South; it does not have influence over the rich industrialized nations.

Herman Daly's perspective enables us to see the distortions built into conventional economics and explains why the current thrust toward economic globalism is so dangerous. Here's another proposal in his parting shots: "Maximize the productivity of natural capital in the short run and invest in increasing its supply in the long run." Any economist knows that when there is a factor that limits production, we should make maximum use of what we have and invest in increasing its supply. But many economists don't look on the "natural capital" that comes from the Earth as a limiting factor.

Economics is based on the notion that human inventiveness can overcome any limits in nature, creating more or finding substitutes. Daly knows this assumption is a mistake. "In the past, natural capital has been treated as superabundant and priced at zero ... Now remaining natural capital appears to be both scarce and complementary, and therefore limiting. For example, the fish catch is limited not by the number of fishing boats, but by the remaining populations of fish. Cut timber is limited not by the number of sawmills, but by the remaining standing forests.... The atmosphere's capacity to serve as a sink for CO_2 is likely to be even more limiting to the rate at which petroleum can be burned than is the source limit of remaining oil in the ground."

But since, by definition, natural capital cannot be made by us, how do we invest in protecting it or in increasing it? According to Daly, we can do this by encouraging different kinds of policies, such as "following investments, allowing this year's growth increment to be added to next year's growing stock rather than consuming it. For nonrenewables ... how fast do we liquidate? ... how much of the correctly counted income do we then consume and how much do we invest? ... The failure to charge user cost on natural capital depletion surely biases investment away from replenishing resources."

Finally, Daly makes the radical suggestion that we "reverse direction from globalization to more national economies. It will not be easy because, at the present time, global interdependence is celebrated as a self-evident good. The royal road to development, peace, and harmony is thought to be the unrelenting conquest of each nation's market by all other nations.... [T]he word 'nationalist' has come to be pejorative."

Daly points out the strange fact that "the World Bank exists to serve the interests of its members, which are nation states, national communities, not individuals, not corporations, not even NGOs. It has no charter to serve the one-world-without-borders cosmopolitan vision of global integration."

We often see this destructive impact of the global economy in Canada and the U.S., where logging, fishing, or mining policies that may foster sustainable communities and ecosystems are opposed by transnational corporations whose goal is to maximize returns for investors. When international agreements are made for global environmental problems, national governments must be able to enforce them. As Daly says, "If nations have no control over their borders, they are in a poor position to enforce national laws, including those necessary to secure compliance with international treaties."

Daly also makes the daring warning that "Cosmopolitan globalism weakens ... the power of national and subnational communities, while strengthening the relative power of transnational corporations.... It will be necessary to make capital less global and more national.... [T]en years from now the buzz words will be 'renationalization of capital' ... not the current shibboleths of export-led growth stimulated by whatever adjustments are necessary to increase global competitiveness. 'Global competitiveness' usually reflects ... a standards-lowering competition to reduce wages, externalize environmental and social costs and export natural capital at low prices while calling it income."

Daly forces us to question our current economic paradigm. He brings a welcome dose of common sense to counter the barrage of mindless repetition of the overriding importance of globalization.

The Wall Street Journal*'s* *Insane Criteria*

I'VE LONG ARGUED THAT GLOBAL ECONOMICS UNDERLIES THE CURrent wave of ecological destruction and must be altered if we are to have a sustainable future. On January 12, 1995, in an editorial headlined "Bankrupt Canada?" the *Wall Street Journal* declared that Canada "has now become an honorary member of the Third World in the unmanageability of its debt problem." It warned that without heroic measures, Canada might have to call in the International Monetary Fund to stabilize its falling currency.

In fact, Canada's resemblance to less-developed countries extends beyond the deficit. We manufacture very few of the products used extensively in our homes, garages, and workplaces. Most of our major industries are branch plants of transnational corporations whose head offices are located elsewhere in the world. Nevertheless, we live like an advanced nation and finance the illusion with income generated by exporting raw materials.

The patronizing attitude toward Third World countries implicit in the *Wall Street Journal*'s editorial reflects a belief in the superiority of the industrialized world and the global economy. Canada has already lost a triple-A credit rating. The federal government, advises the *Wall Street Journal,* should emulate the provincial premiers who cut rural hospitals, chopped school boards, and slashed welfare rolls.

Before our finance minister follows the prescription of the business community and the *Wall Street Journal,* let's try a suggestion once made to me by economist Kenneth Boulding. I was describing the scary rhetoric used to "persuade" Canadians to adopt the Free Trade Agreement (FTA) with the United States. Without the FTA, we were being told, Canada would be an "imminent economic basketcase." Boulding responded that to get an idea of how rich we really are, just imagine the consequences for Canada if all other countries vanished, leaving only our country and two hundred miles of ocean. What would happen?

Canada is one of the two major grain exporters in the world, so we would not lack for food. Our natural resources of land, water, timber, and minerals make us the envy of most people on the planet. Furthermore, we have an educated, literate public, capable of creating the most advanced technology desired. Boulding's thought exercise reveals that we are extremely well off.

If we apply the same exercise to Japan, long considered one of the world's economic giants, and China, now roaring into economic prominence, the result is very different. Either country, if left alone, would be in immediate difficulty. Japan has little energy and few natural resources beyond trees, and its population is able to live high on the ecological food chain only by exploiting waters and resources far beyond its borders. China, in contrast, does have energy but also has major problems with pollution and limited agricultural land and resources. Each country on its own would be plunged into desperate straits.

There is something absolutely perverted in a system that identifies Canada as a poor country requiring external economic institutions to stabilize its economy, while Japan and China are considered economic colossi. Global economics grossly undervalues the air, water, soil, and biodiversity that keep us alive while vastly overvaluing human intellect and activity. That's what creates the dichotomy in wealth between Canada, Japan, and China.

On the same day the *Wall Street Journal* was making its pronouncements, U.S. president Bill Clinton declared American confidence in the Mexican economy to help stabilize the calamitous plunge in the value of the peso. For weeks the Mexican government tried frantically to stabilize its currency. Pundits declared that the Mexican economy had not inspired "investor confidence."

Even if we overlook the glowing promise that seduced Canada and Mexico into the North American Free Trade Agreement (NAFTA), what does the peso's plummet reveal? It informs us that currency speculators are now more powerful than governments. The Mexican government dipped deeply into its reserves to buy up pesos, all in a vain effort to halt the peso's slide, just as the French government and banks had done two summers earlier to halt the plunge of the franc. Few governments can stand up against more than U.S. $600 billion in daily currency speculation on stock markets around the world.

It is an insane game in which money can be bought and sold to make more money. In this kind of game, money can be created faster than things in the real world like food, fish, or trees, and the players don't have to worry about long-term sustainability. Currency is a human invention, and when it is no longer anchored to anything more tangible than "investor confidence," we enter an Alice-in-Wonderland fantasy world. It is a realm that relegates Canada to Third Worldhood and places little value on the future of communities or ecosystems.

Following a Different Path

POLITICS IS A DOMINANT FORCE IN OUR SOCIETY. THE VALUES AND actions of our elected representatives lead us into the future. We think this is the only way things can be, but there are societies with very different ways of governing themselves, and they clash with what we are used to.

Having been adopted by the Eagle clan of the Haida people, I attend the annual assembly of the Haida nation in Haida Gwaii (Queen Charlotte Islands). At the end of the meetings, we all take part in a wonderful feast. Tables are heaped with homemade bannock, cakes, and pies, along with clams, crab, herring roe, salmon, and halibut caught in local waters. After eating, guests are showered with gifts and treated to songs and dances.

Here, culture, history, and community are rooted in the abundance of the forests and surrounding waters that define "home" and "belonging." The Haida bring to mind a remarkable speech delivered in 1992 before the United Nations by Anderson Mutang Urud, a Kelabit from Sarawak in Malaysia. His words were rich with ideas and insight for all of us.

While Sarawak is less than 2 percent the size of Brazil, it produces almost half of the world's tropical timber. Even if the rate of logging is cut in half immediately, Mutang warned most of the primary forest in Sarawak would be gone by the year 2000. And as the forest is cleared, there is a domino effect: "Fish, wild animals, sago palms, rattan, and medicinal plants disappear. The trees which bear the fruit which feeds the wild pigs are cut down for timber. The pigs disappear, and with them vanishes the main supply of meat for our peoples.... Trees and vines with poisonous barks are felled, and find their way into the streams, killing all the fish. Mud from the eroded lands pollutes the rivers, bringing us diseases and destroying our source of drinking water."

Like Canadians, Malaysians have shown little respect for the sacred burial sites of the indigenous people: "The logging companies bulldoze through them, with no regard for our feelings.... When we complain about their destruction, they sometimes offer us a small sum of money as compensation, but this is an insult to us. How can we accept money that is traded for the bodies of our ancestors?"

There is a fundamental clash between value systems. *Progress* and *development* are the words government uses to justify the logging activity. To this, Mutang says: "For us, their so-called progress means only starvation, dependence, helplessness, the destruction of our culture and demoralization of our people."

Like those of British Columbia, the forestry practices of Sarawak are aimed at maximizing profit with little regard for sustaining communities or conserving the forests. Mutang sees this as the ruse it is: "The government says it is creating jobs for our people. But these jobs will disappear along with the forest. In ten years, the jobs will be gone; and the forest which has sustained us for thousands of years will be gone with them."

The real problem is that Western notions of progress tear apart ways of life that support communities and families: "My father, my grandfather did not have to ask the government for jobs. They were never unemployed. They lived from the land and from the forest.... we were never hungry or in need. These company jobs take men away from their families for months at a time. They are breaking apart the vital links which have held our families and our communities together for generations. These jobs bring our people into a consumer economy for which they are not prepared.... The Penan, the Kelabit, and the other indigenous peoples view the forest as our home. When we see a thief enter our home, we try to defend what is ours." Since 1987, there have been peaceful blockades to which the government has responded by arresting and imprisoning scores of people. Mutang himself has been tried in absentia and will be arrested if he ever returns to the country.

Mutang relates:

> A high government official once told me that in order to have development, someone must make a sacrifice. I replied, why should it be us who must make this sacrifice? We have already become poor and marginalized ... while companies get rich from our forests, we are condemned to live in poverty. Now there is nothing left for us to sacrifice except our lives.
>
> In our race to modernize, we must respect the ancient cultures and traditions of our indigenous peoples. We must not blindly follow that model of progress invented by European civilization. We may envy the industrialized world for its wealth; but we must not forget that this wealth was bought at a very high price. The rich world suffers from so much stress, pollution, violence, poverty, and spiritual emptiness. The wealth of indigenous communities lies not in money or commodities, but in community, tradition, and a sense of belonging to a special place.... The world is rushing toward a single culture. We should pause and reflect on the beauty of diversity.

Mutang's appeal on behalf of his people is relevant here in Canada. In Haida Gwaii, the land and waters are still rich with life that gives an opportunity to define progress differently and follow a different path.

Update

Environmentalists have long criticized Malaysia and the government of Sarawak for logging practices that are both unsustainable and destructive pof the forest home of the Penan people. Malaysia showed little concern in the face of global condemnation. Now, in quick succession, we have learned of a massive cloud of smoke that hung over Indonesia and Malaysia for weeks in 1997, while in 1998, reports have revealed that drought-induced fires have been burning uncontrollably in Sarawak. Pictures of suffering orangutans have replaced the images of the Penan as symbols of the victims of avaricious forest destruction.

The True Price of a Tree

WE WERE STANDING IN AN ANCIENT FOREST THAT WAS THREATENED with clear-cut logging. He was the CEO of the company that had been allotted the tree farm license that enabled him to drive a road into the valley and begin the industrial extraction of the trees that would destroy what took millennia to evolve. We had engaged in an animated argument about the fate of that forest while standing face to face less than half a meter apart.

"You see that tree over there?" he shouted, pointing to a giant that was probably many centuries old. Without waiting for my response, he continued, "It doesn't have any value until it's cut down." I was dumbstruck by the statement, giving him the opening to carry on. "Unless, of course, you tree huggers decide you'll pay money to save it so you can enjoy it. Think your cronies can raise enough money to save the entire forest? Logging is what keeps the economy of this province growing and makes it possible for you preservationists to wear clothes, drive cars, and watch TV."

What had made me speechless was the realization that he was right. In the value system inherent in the form of economics our society has embraced, only when money is exchanged for goods and services is the transaction recognized as having economic worth. But the perspective through which I viewed the forest of which that tree was a part was radically different. That one tree was a tiny part of a community of organisms thousands of years in the making. That community is made up of trees ("merchantable timber," or "fiber," in the jargon of the industry) that are a tiny minority of the life-forms that comprise the forest. The soil is a living organism made up of tens of thousands of species of microorganisms—viruses, bacteria, fungi, protozoa—and larger nematodes, worms, insects, and mites. Plants and animals blanket the forest floor, lichens and mosses coat rocks and decaying wood, snags and fallen logs provide nutrition and protection for countless organisms. This is the community that we recognize as a forest, complex and interlinked beyond comprehension and all held together by the air, water, and sunlight that suffuse them.

What foresters refer to as "second-growth forest" is not a forest at all but a tree plantation, an attempt to grow trees like a crop of tomato or corn plants. But of course, such a "managed forest," or "fiber farm," no longer has the resilience, regenerative capacity, or self-protective devices of a natural forest, and so companies require geneticists to breed fast-growing strains of commercially valuable trees, tree planters, herbicides to clear out "weed" (i.e., commercially worthless) species, insecticides to eliminate pests, fertilizers to restore nutrition to soil, and firefighters. Large clear-cuts and use of heavy machinery expose soil flora and fauna to sunlight, wind, and air, alter water retention and streams and rivers that are the lifeblood of the forest, and radically transform the species mix left.

Foresters rationalize these practices as "proper silvicultural management," as if they know what it is that creates the original forest that was cut down. Of course, they have no idea. They lack both a proper inventory of all the constituent species that make up the forest and a blueprint that explains how all of the components are interconnected. The entire forest is like the goose that laid golden eggs in the children's fable; as long as the goose is fat and healthy, it will yield golden eggs indefinitely. In the short-term perspective of global economics, as in the children's story, forestry companies attempt to gather all of the eggs at once by killing the goose. As I pointed out in the book *Good News for a Change: Hope for a Troubled Planet,* by selective logging at or below the growth rate of the trees in a forest, trees can be profitably "harvested" indefinitely instead of once every hundred years or more. Nor is the diversity that is the key to resilience and regeneration sacrificed when trees are selectively removed.

Returning to the CEO's statement that the tree only acquires value when it enters our economy by generating revenue, consider what the tree does before humans define its value. Hundreds of years old, that tree has absorbed carbon dioxide (a greenhouse gas) from the air, thereby playing a part in life's climate engine, and releases oxygen as a byproduct of photosynthesis (not a bad byproduct for all animals like us that are completely dependent on that oxygen for survival). The energy of photons absorbed by the tree's leaves is transformed into sugar molecules, which, like fossil fuels, store that energy to be liberated in the controlled reactions of metabolism. The roots of the tree cling to soil even under the hardest of rains, thereby inhibiting erosion while siphoning vast quantities of water up into the canopy, where it is released through transpiration, and hence ameliorating weather. From roots to tips of branches, the tree offers a habitat for countless forms of life, from lichens and fungi to insects, birds, and mammals. All of these "natural services" performed by that standing tree affect human health and survival but are ignored by our economy. It's long past time when we started lifting our horizons and values beyond the extremely limited perspective of conventional economics.

Plundering the Seas

EARTH IS REALLY A WATER PLANET. AS LAND DWELLERS, WE ARE AIR chauvinists who are immensely ignorant of what lies beneath the ocean waves. Yet while atmosphere, climate, and topsoil are being altered, marine systems seem to be buffered from rapid change. So, increasingly, we are turning to the oceans for resources, especially food.

The oceans cover 70 percent of the planet's surface. They are where life itself began, and they are home for a vast array of living things. Today, those waters are being plundered on a scale that is catastrophic. If we protest the burning, clear-cut logging, and damming of the Amazon rain forest, British Columbia's coastal rain forests, California's redwoods, and the James Bay watersheds, we cannot ignore the ecological devastation being inflicted on the oceans.

We are taking too much and destroying the oceans' habitats in the process. You can get a hint of the cause of the problem by visiting Tokyo's Tsukiji Fish Market, the largest in the world. Kilometers of tightly packed aisles are crowded with merchants selling sea life in mind-boggling volume and variety. Row upon row of huge frozen tuna, swordfish, and sharks are sold on the docks. Fish eggs of many species, tiny fish fry, small octopuses and crabs, slabs of deep red whale meat and fish of every size, shape, and variety are for sale. By afternoon, the stalls are empty, and the next morning, the process is repeated. Since Japan's insatiable appetite cannot be satisfied from its own waters, it buys or catches food from the oceans of the world.

Everything about the causes of the environmental crisis—ignorance, shortsightedness, greed—is exemplified by the way we are "harvesting" the oceans. Our knowledge of ecosystems in the aqueous world is extremely limited. So we act as if things that are out of sight needn't be thought about. How else can we explain the way we use the oceans as dumps for garbage, sewage, industrial effluent, nuclear waste, and old chemical weapons?

Moreover, our fishing policies are determined by political and economic priorities instead of the requirements of complex marine ecosystems. Japan's fishing fleet plies the Seven Seas like twentieth-century buccaneers, looting and pillaging "international" waters with impunity. It is the drift nets of Japan, as well as Taiwan and Korea, that represent the ultimate in greed and shortsightedness. If we were willing to go to war over Kuwait, we should respond just as massively to the vandalism in the seas.

Common sense ought to inform us that when 50,000 kilometers (30,000 miles) of near-invisible nylon nets are set nightly six months a year, the destruction will be unacceptable. These "curtains of death" form walls that trap far more than the squid they are said to be fishing—they indiscriminately catch fish, marine birds, and mammals, including porpoises, seals, and small whales. Like deadly scythes, drift nets cut broad swaths through the ocean. Fishers use drift nets as if the animals caught will be somehow endlessly replaced. Japan's assurances that the effects of drift nets will be "monitored" are a cynical ploy. This ocean strip-mining should be stopped immediately.

The oceans are vast and often treacherous, and each season, hundreds if not thousands of kilometers of drift nets are lost. No longer under human control, they become "ghost nets" and continue to catch fish around the clock all year. When they acquire a heavy load of victims, the nets may sink to the ocean floor until the carcasses decay and fall off. Then, like an invisible undersea monster, the nets rise again to continue their deadly chore.

The long-term ecological effects of our marine activity have to be rated above the short-term economic benefits. And it's not just drift nets—we treat the ocean floor as if it is uniform and endlessly resilient. Immense factory ships drag massive weights and nets across the ocean bottom in an ecologically destructive operation that is like clear-cutting a forest. Large clams called geoducks, many more than a hundred years old, are blasted out of the ocean floor with jets of water that destroy habitat for many bottom dwellers.

Governments assume that by regulating catch limits, fishing gear, and fishing seasons, ocean resources are "managed." Instead of beginning from a sense of our vast ignorance of ocean ecosystems and productivity and designing regulations that always err on the side of caution, we allow organisms to be taken until they disappear. If we mean it when we talk about a "sustainable future," we have to change the way we exploit the oceans. Drift nets are a place to start.

Update

After Japan had created drift nets made of deadly, efficient, and tough nylon monofilament in the late 1970s, the number of boats using them increased enormously. Used primarily to catch squid, the nets were so effective that Japan banned them within 1,000 kilometers (600 miles) of its shores, thereby sending Japanese boats to fish far out at sea. More than five hundred boats in the Bering Sea in the 1980s, mainly from Japan and Taiwan, led Alaska to expel the Japanese fleet from the Bering Sea and led New Zealand to ban Taiwanese drift nets from their waters in 1983. In 1989, Hawaii banned them, South Pacific nations signed the Tarawa Declaration calling for a ban, and the Canadian minister of Fisheries and Oceans, Tom Siddon, declared his opposition. The FAO (Food and Agriculture Organization) estimated that in 1989, between 315,000 and a million dolphins were killed in drift nets. In December 1989, the United States introduced a resolution in the United Nations calling for a ban on drift nets in the South Pacific and a worldwide moratorium by June 30, 1992. In order to pass the resolution, it was watered down.

Observers aboard fishing boats led to estimates that Japanese drift nets took 106 million squid and 40 million others of a hundred species. On November 25, 1991, Japan and the U.S. announced an agreement to phase out Japanese drift nets throughout the world. On December 20, 1991, the UN passed a resolution for a 50 percent reduction by June 30, 1992, and a full phaseout by the end of 1992. European boats continued to use drift nets in the Atlantic and the Mediterranean until the late 1990s.

Drift nets had scythed a terrible swath through fish, mammals, turtles, and birds by the time the nets were phased out. But other deadly methods were already in use. Longlines that could stretch up to 50 kilometers (30 miles) and carry 10,000 hooks were and still are being deployed throughout the oceans. As they are played out, the lines extend close to the surface far behind the ships, where the baited hooks attract seabirds, especially albatross. Between 1997 and 2000 in the South Pacific alone, it is estimated that 330,000 seabirds, including 46,500 albatross, 7,200 giant petrels, and 138,000 white-chinned petrels, were drowned by longlines. In addition, the longlines take sharks, turtles, dolphins, and even whales, which become entangled. In 1994, the bycatch (nontarget species) with longlines off Alaska alone was 572 million pounds.

If that weren't enough, technological improvements allowed immense nets dragged along the ocean bottom by two trawlers to haul up anything in their way. Nets made of strong, synthetic fibers are big enough to hold twelve jumbo jets! In six weeks, French and Scottish pair trawlers caught more than two thousand dolphins. Not only are huge amounts of bycatch taken on board and then discarded, the weights to hold the nets down are immense rollers that crush the habitat of the sea bottom. Marine conservationist Elliott Norse estimates the total area of sea bottom dragged annually is 115 times as great as the area of forest clear-cut throughout the world each year.

The crisis is in the development of powerful technologies that can increase yields dramatically but destroy habitat, kill a wide spectrum of noncommercial species, and increase the take of the economically valuable targets. Without ecological principles of sustainability to limit their take, new techniques fuel a gold-rush mentality without regard to the future.

Shifting Political Perspectives

IN MARCH 1991, THE BRITISH COLUMBIA FOREST MINISTRY SPONsored a conference on biodiversity. Various experts discussed the values of biodiversity in forests and watersheds and the threats posed by human activities such as logging, mining, dams, pollution, and agriculture. After the opening speech, a member of the audience rose and spoke emotionally of his fears and frustration. He told us he had dropped out as a hippie in the 1960s and returned as an environmental activist in the 1980s. But accelerating destruction of the forests was pushing him to consider ever more desperate measures, including violence and breaking the law. Would we join him? he challenged us.

The power of the environmental movement has been its uncompromising commitment to nonviolence. But increasingly, people sense the urgency of the global environmental crisis and express rage at and frustration with government reticence or inaction.

The history of chlorofluorocarbons (CFCs) is instructive. When freon, a kind of CFC, was announced in 1930, it was hailed as a wonder chemical—stable, nontoxic, nonflammable—from the folks who had brought us nylon, plastic, and DDT. By 1976, over 750 billion pounds were being made in the United States alone for use as propellants in spray cans, refrigerants, and cleaning solvents for silicon chips. In the early 1970s, scientists discovered that in the upper atmosphere ultraviolet radiation (UV) breaks chlorine off CFC molecules and the liberated chlorine is highly reactive, breaking down ozone.

Ozone in the upper atmosphere absorbs most of the UV from the sun. The small amount getting through induces a tan, but in heavier doses, UV causes genetic damage to the DNA itself. It kills microorganisms, in people induces skin cancers (including deadly melanoma) and cataracts and reduces the effectiveness of the immune system. We don't know what the increasing intensity of UV will do to forests, ocean plankton, agricultural crops, or wildlife.

In response to the scientific announcements, consumer demand for spray cans declined, thereby pushing the chemical industry to develop alternatives containing less chlorine. The issue then died down until 1984, when atmospheric scientists reported an immense "hole" in the ozone layer above Antarctica. This discovery galvanized the scientific, political, and industrial communities as well as the public. By 1990, more than fifty countries had agreed to a total ban on CFCs by the year 2000.

Few environmental issues are as clear-cut as the CFC-ozone story. The offending causative agent was identified, the effects on ozone and UV are known, and less destructive alternatives are available. Yet there were too many uses for CFCs that had become too important economically to stop their production immediately. Even after they are banned, CFCs will continue to escape into the air for decades. Compared with ozone depletion by CFCs, ecological problems such as global warming, toxic pollution, or rain forest destruction are far more complex, thus making it extremely difficult to determine and quantify cause, effects, and solutions. If it is taking so long to act on CFCs, how much longer will it take to respond to other issues?

The environmental impact of massive dams, clear-cutting, and toxic pollution is already so serious that immediate and heroic measures are needed. In one year alone, citizens took the Canadian federal environment ministry to court to try to force the government to do what it is supposed to, namely, carry out environmental impact studies of the Kemano Completion Project and the logging of Clayoquot Sound in British Columbia, the Oldman River Dam in Alberta, the Rafferty–Alameda Dams in Saskatchewan, the James Bay hydroelectric project in Québec, and the Point Aconi coal plant in Nova Scotia.

It's grotesque that the public has to take our elected representatives to court to make them do what they are, by law, supposed to do. Is the answer to elect the right kind of politicians or party? Or are there fundamental flaws in the structures and priorities of government itself that regardless of politician or party preclude serious action on major environmental issues?

Any form of government is a human creation and reflects the value and belief systems of a people. And like all human creations, no political system is perfect. In times of stress and rapid change, as we have seen from the former Soviet Union to South Africa, governments require flexibility, imagination, and leadership. These qualities become especially critical as the planet's biosphere undergoes unprecedented change.

Canada is a young country whose system of democracy is still in flux. Canadian-born Japanese like my parents were only given the right to vote in 1947, and First Nations people received the same right in 1960.

Nowadays, there is a pervasive public cynicism toward politics and politicians that is an expression of frustration with the apparent inability or unwillingness of politicians to act and their often terribly slow response when they do. But maybe government priorities and perspectives can't be congruent with the needs of the environment. Maybe even the best of people with the highest intentions, the right political priorities, and best government structures would be unable to take adequate action on global ecological degradation.

Since politicians act on behalf of their constituents, the human beings who elect them, they are concerned with other life-forms only as they impinge on human needs and demands. This narrow focus may not have mattered in the past, but now that we have acquired the capacity to wipe out entire ecosystems almost overnight, this species chauvinism blinds us to what we are doing to the rest of nature. At the very least, all people must acquire an understanding of ecology and recognize that we remain embedded in and dependent on an intact natural world.

Consensus is necessary to avoid "playing politics" with the fate of the planet. Today, politicians must make decisions about issues ranging from soil degradation to toxic pollution, atmospheric change, deforestation, species extinction, environmental carcinogens, and fusion power, issues that require a grasp of fundamental principles, concepts, and terminology in science and technology. Yet most of the elected officials in Canada and the U.S. come from two professions—law and business—that are ill-equipped to deal with science and technology. The disproportionate representation from business and law skews government priorities so that economic matters like free trade, softwood lumber, and the GST and jurisdictional issues like Canadian sovereignty in the Arctic and Québec independence dominate the political agenda. Our education system must ensure that *all* people are scientifically and technologically literate so that they understand and respond to crises in these areas.

Each of these obstacles can be overcome, but to do so requires a radical shift in the societal values, assumptions, and beliefs that shape political perspectives.

Lessons from Humanity's Birthplace

BY COMPARING DNA FROM DIFFERENT SOURCES, SCIENTISTS HAVE concluded that the common ancestor of all of humanity may have lived along the Rift Valley of Africa, one of Earth's landmarks that is visible to astronauts in space.

Here in the cradle of humanity at the north end of the fabled Serengeti Plain, we get a hint of what the conditions may have been like when those first human creatures made their appearance. Descending from the trees and walking upright, they shared a vast expanse of grassland with herds of grazing and gamboling animals beyond count.

In Masai Mara, as in the Galápagos Islands, the animals haven't learned to fear our species. Large gatherings of wildebeest graze along with Grant's and Thompson's gazelles, zebras, and topi. We encounter giraffes, elephants, and lions along the road's edge. The Mara River is filled with hippos bellowing and blowing, while immense crocodiles lie motionless on the banks.

For the bird lover, this is truly a paradise, as flocks of dignified crested cranes glide by, plovers leap screaming at our intrusion, and even starlings are spectacular. There are frogs, lizards, and snakes to satisfy the most demanding herpetology buff.

As a longtime lover of insects, I am enthralled by their variety here, from praying mantis to walking sticks, grasshoppers as big as mice, and scarab beetles rolling balls of dung. (When I was a child, I avidly captured and killed insects for my collection, but how times have changed. My daughters share my passion for insects but adamantly refuse to kill them just for display. Now pictures have replaced boxes of pinned specimens.)

For city dwellers from industrialized countries, the Serengeti's diversity and abundance of wildlife are a stunning contrast to our barren urban habitat. But coming from such an impoverished environment, we have no basis for judgment about the state of life here on the plain. To get a fuller perspective, we read records of early adventurers and talk to elders, men and women who have lived here for their entire lives. They inform us that the makeup and numbers of animals have been profoundly altered in this century, and that in spite of greater awareness, national parks, and ecotourism, the change continues.

Only a few decades ago, one of the truly charismatic mammals in the world, the rhinoceros, numbered in the tens of thousands. Today in Masai Mara, there are fewer than twenty, and African children now take for granted a profoundly restricted range and number of their animals. Cheetahs, chimpanzees, and leopards that once flourished across large parts of the continent are currently confined to tiny areas in parks, reserves, and private game farms.

When Ed Sadd, the owner of Bushbuck Camp, where we are staying, came to Nairobi in 1978, Kenya's population was 12 million people. Today it is 28 million. Kenya has one of the highest birth rates in Africa, and its population will double again in a mere 17½ years. This increase in population puts enormous pressure on the habitat of the wild animals. It is simply not enough to set aside tracts of wilderness for animals when exploding numbers of people need income, land, wood, and food. Somehow the needs of both have to be satisfied and protected.

Like millions of others from industrialized countries, I am here as a tourist. "Ecotourism" is much touted as a way for countries in the South to preserve wilderness while generating much-needed income. But tourism is not benign; it too is "consumptive." Here on the Mara, the impact of tourism is readily apparent.

It has rained heavily and the ground is saturated with water, so the four-wheel-drive vehicles quickly cut crisscrossing tracks and ruts across the plains. Trucks and vans frequently bog down, their spinning wheels cutting deep trenches through the grass.

We have chosen to stay put in one camp and take our time seeking and observing specific species. But most tourists stay in this area for only two or three days, so their drivers whip across the grass to maximize the number of animals they can see.

The drivers watch other vehicles closely. We come upon a male lion mating with two females, and soon there are eight other cars surrounding the animals and parked within a few meters of them. It is a thrilling experience to have such an intimate view of nature, and yet the snickering and giggling of the tourists and our close proximity make it seem as if the animals are performing for our titillation.

There are few places on Earth that have the number and diversity of animals that still live here in Kenya and Tanzania. But they face unprecedented challenges of pressure from human beings. Can we learn to share their space so they can flourish?

Ecotourism is certainly preferable to extensive logging, damming, or otherwise developing wilderness. But ecotourism is invasive and consumptive. Ultimately, we will have to learn to revel in knowing that wildlife exists free and beyond our presence.

True Wealth

CHIEF SKIDEGATE OF THE HAIDA PEOPLE HAD TAKEN MY FAMILY AND me on his boat to explore the remnants of a village on the west coast of his land, Haida Gwaii. With hushed voices, we walked past fallen beams of a great longhouse, moss-blanketed outlines of the village, and weathered poles that gave mute testimony to an ancient culture. Later we caught salmon and lingcod and gathered rock scallops, sea cucumbers, and sea urchins to eat. Sitting around a bonfire, groaning with overstuffed stomachs, we drummed and sang Haida songs. Leaning back against a log, his belt undone to accommodate his full gut, Chief Skidegate reflected, "I wonder what the poor people are doing tonight." We all chuckled with recognition that it can't get much better than that, and we really were the wealthiest people in the world. How could we put a price on a day like that?

Last January I flew into Aukre, a Kaiapo village deep in the Amazon rain forest. There the three hundred inhabitants live as traditional a life as I can imagine. They are twenty-first-century people, now linked to the rest of the world by their airstrip and solar-driven radiophones and dependent on metal machetes and axes, rice and sugar, medicines, and some clothing from what they call "civilization." But they have deliberately chosen to live traditionally, resisting the call to settle in towns and relying instead on the energy of the sun and fire, water running in the river, and their forest, which is pharmacy, grocery, and lumberyard.

My friend of fourteen years, Chief Paiakan, greeted me like family. Knowing my love of fishing, he took me onto the river the next day. It was the rainy season, and the swollen river had flooded the forest. It had been totally different on my last visit, when the river was low and we had sought pools for electric eels, tucunare, and giant catfish while searching the banks for turtle eggs. This year the river was a torrent and fish swam among the trees in search of fruits and nuts. Casting hooks embedded in fruit, we caught a variety of fruit-eating fish. I remarked to Paiakan, "You know, people work hard for years to save up enough money to go on a holiday to do this." He stared at me with incomprehension.

Somehow we have forgotten what really matters. We used to say, "The best things in life are free." Now it seems to be "The best things in life cost the most." Last June in Washington, I gave a speech at the World Bank, an institution that is harshly judged by social justice and environmental activists. In several programs, I've pointed out the World Bank's shameful environmental record, so I was delighted to have the opportunity to address the bank directly. I was frank and tough in my criticisms but offered an avenue for dialogue and change. (To my amazement, I was rewarded with a standing ovation.)

During the question period, an economist challenged, "You have pointed out where changes have to be made, but instead of criticizing, why don't you work with us? It's just a matter of getting the pricing right." And there we had it. Nothing could be clearer on the problem with the kind of economics our business and political leaders have embraced and are pushing around the world. "It's just a matter of getting the pricing right," the implication being that everything has a price, right? Wrong. The most important things are beyond price. How much is your mother worth? Or your sister or your child?

I once received a letter from a real estate agent when offshore investors were buying up Vancouver real estate. The letter suggested this was an ideal time to sell my house and "buy up." After twenty-five years of living in this house, it was more than a piece of property, it had become my *home.* If I were to make a list of those things that mattered to me and made it a home, I would include:

- The cemetery under the dogwood tree, where our family dog, Pasha, was buried along with Blackie the cat and snakes, mice, and birds the children had found on the street
- The asparagus and raspberry patches planted just for me by my father-in-law, who lives upstairs
- The cabinet that my father had built for our first apartment when Tara and I were married and that I had installed in our kitchen when we bought the house
- The handle carved for the back gate by my best friend, Jim Murray, when he came out and stayed with us for a week
- The clematis plant along the fence where we put some of my mother's ashes when she died

Those things on that list are what transformed a house into my home and make it priceless beyond anything monetary. But in the marketplace, they are worthless.

The Midwestern farmer who refuses to let the bank claim his land because his grandfather had cleared it, his father passed it on to him, and he intends to pass it on to his children and grandchildren reflects similar values. The Newfoundlanders living in remote outports who resist the allure of electricity and running water to remain where their ancestors settled centuries ago attest to an attachment to place that transcends economics. Then consider the value system of First Nations people, who have occupied land for millennia, who regard the animals and plants as their kin, and for whom rocks, rivers, and forests are sacred.

And that, it seems to me, is the problem with the mentality that suggests that "it's just a matter of getting the pricing right." In mistakenly equating money, and the plethora of products generated by an economy built on consumption, with wealth, we have lost sight of those things in life that really matter most to us and are worth far more than anything money can buy.

SCIENCE, TECHNOLOGY, AND INFORMATION

THE TWENTIETH CENTURY HAS BEEN DOMINATED BY SPECTACULAR insights acquired by scientists and by the application of many of those ideas as technology. When I was born, in 1936, there was no need to worry that children were watching too much television, because it had yet to become widespread in homes. Back then, smallpox killed millions and polio was a feared disease. When I was a child, there were no jets, videotapes, CDs, satellites, transoceanic phone links, computers, oral contraceptives, CAT scans, genetic engineering, or organ transplants, among many other things. Each innovation changed the way we lived and the way we thought of ourselves. A near-endless stream of new consumer products titillated the consumer.

Science has come to occupy a position of omnipotence, pushing back the curtains of ignorance and revealing answers to nature's deepest secrets. We began to believe that with greater discoveries, we would gain the knowledge needed to understand and control the forces impinging on us and life for all would continue to improve.

As the time period between discovery and application shrank, the distinction between basic research and development of technology was often difficult to delineate. And as pure scientists saw the usefulness of their work, many became caught up in the rush to use every insight for profit, power, or benefit. There is much to learn from the history of science and technology of the twentieth century.

Science and Technology Are Still in Their Infancy

ONE OF THE MAJOR UNDERLYING CAUSES OF THE GLOBAL ECOCRISIS is the dominant attitude within society today. It is based on a faith in the power of science and technology to give us insight and understanding that enable us to control and manipulate our environment.

Technology has revolutionized human evolution since the earliest records of our species. While providing practical dividends in the past, technology—whether pottery, painting, bow and arrow, or metalwork—did not require scientific explanation. But today, it is science that drives technological innovation, from telecommunications to biotechnology and nuclear power. And now, our insights and inventions have given our species unprecedented power to change our surroundings with unpredictable consequences. For this reason, it is essential to understand the nature of the scientific enterprise, what it reveals, and where its limits lie.

As a brand-new assistant professor in the early 1960s, I taught a course in genetics, my field of specialty, with all of the enthusiasm of an ambitious hotshot on the ladder to tenure, recognition, and bigger grants. After one of my first students had been in my lab a while, he remarked that he had assumed that geneticists knew almost everything. "Now that I've been doing experiments for a year," he went on, "I realize we know almost nothing."

He was absolutely right, of course. In spite of the vaunted "success" of modern science, it has a terrible weakness, one that is inherent in its methodology. Scientists focus on a part of the world that they then isolate, control, and measure. They gain an understanding of and power over that fragment of nature without knowing how it meshes with other components of a system. The insights we acquire are a fractured mosaic of bits and pieces instead of an integrated whole. Thus, we may invent powerful techniques to manipulate genetic material, for example, without knowing what it will do to the whole animal.

Those who have been practicing science for a while know that experiments are far more likely to yield a puzzle than a satisfying answer. So while the spectacular pictures from satellites passing by our neighboring planets may have eliminated a few theories, they generated far more questions. There is something reassuring in knowing that nature is a lot more complex than we can imagine.

But the practice of science has changed radically during the past decades. After *Sputnik I* was launched in 1957, the United States responded by pouring money into universities and students to catch up. It was a golden period for scientists as good research in just about any area was supported. When I graduated in 1961 as an expert on the behavior of chromosomes in fruit flies, my peers and I could choose from several job offers and grants. We were engaged in a quest for knowledge purely for the sake of knowing, and we took it for granted that good research would eventually lead to ideas that could be applied.

In the ensuing years, science has become extremely competitive because of the high stakes that come with success. Thirty years ago, a productive scientist in my field might publish one major paper a year. Now several articles are expected annually, and a publication record of a dozen or more is not unusual. But today's articles are often repetitive or report small, incremental additions of knowledge, thereby fragmenting knowledge even further.

Since scientific ideas and techniques have created spectacular new high-tech industries, governments perceive research as vital fuel for the economic engine. Consequently, research funding agencies now look for work that promises to pay off in some practical way, and when applying for grants, scientists have to play a game by claiming or implying that the research being proposed will lead to some beneficial discovery. If you look at the titles of Canadian or U.S. research grant proposals, you would think that all of the world's problems could be solved by scientists right here.

Of course, that's not true at all. Even if we funded people adequately (which Canada does not), few if any of those solutions will be achieved as projected. The game of grant-seeking perpetuates a mistaken notion of how science is done. Scientists do not proceed linearly to a specific goal, going from experiment 1 to 2 to 3 to a cure for cancer, for example. If research worked that way, doing science would be routine and far less interesting. The fact is, most scientists start from an initial curiosity about some aspect of nature. They design experiments to satisfy that interest, then lurch down unexpected side streets, blunder into blind alleys, and, perhaps, through luck and perceptiveness, connect unrelated ideas to produce something useful.

But many young scientists actually believe that science advances in a straight line and that the claims made in grant proposals can be achieved. And the media tend to reinforce the notions with breathless reports of new discoveries and liberal use of the word *breakthrough*. People are relying on this unwarranted optimism when they believe the "experts" will take care of a problem.

But the consequences of the major hazards facing us today—atmospheric change, pollution, deforestation, overpopulation, species extinction, and so on—cannot be scientifically predicted, let alone resolved, because we have only a fragmentary understanding of nature. When scientists say "more information is needed" before a course of action can be planned for an issue like global warming, they give a mistaken impression that such knowledge can be quickly acquired and that, until it is, the problem isn't real, so we can carry on with business as usual.

Scientists who claim their work will solve global hunger, pollution, or overpopulation do not understand the social, economic, religious, and political roots of the problems that preclude scientific solution. There is a vital role for science today in detecting and warning of changes and unpredictable hazards, but scientists have to get rid of the pernicious myth about the potential of their work to solve all our problems.

Biotechnology: A Geneticist's Personal Perspective

Dr. Faustus, Dr. Frankenstein, Dr. Moreau, Dr. Jekyll, Dr. Cyclops, Dr. Caligari, Dr. Strangelove. The scientist who does not face up to the warning in this persistent folklore of mad doctors is himself the worst enemy of science. In these images of our popular culture resides a legitimate public fear of the scientist's stripped down, depersonalized conception of knowledge—a fear that our scientists, well-intentioned and decent men and women all, will go on being titans who create monsters.

—THEODORE ROSZAK

A Personal History

I would like to begin by providing some history and context that may elucidate where my perspective comes from. I was born in Vancouver, British Columbia, in 1936. Both of my parents were born in Vancouver about twenty-five years earlier. In 1942, my Canadian-born and -raised family was stripped of all rights of citizenship, our property and assets were seized and sold at fire sale prices, our bank accounts were frozen and ultimately looted, and we were incarcerated for three years in primitive camps located deep in the Rocky Mountains. Our crime was sharing genes with Canada's enemy, who was also our enemy because we were Canadians. As World War II drew to an end, we were confronted with two choices: renounce citizenship and receive a one-way ticket to Japan or leave British Columbia and resettle east of the Rocky Mountains.

Pearl Harbor and our subsequent evacuation, incarceration, and expulsion shaped the lives and psyche of all Japanese Canadians. For me, those events created my hang-ups (about my slitty eyes) and my drive to prove my worth to fellow Canadians. And the results of the war left me with a lifelong knee-jerk aversion to any hint of bigotry or discrimination and a passion for civil rights.

Falling in Love with Genetics

All my life, nature was my touchstone, my life, my passion. As a boy, my love of fish and fishing led to a hope of becoming an ichthyologist. Later, when my mother sewed me a net for collecting insects, my dreams were transformed to a life of entomology. After moving to Ontario at war's end, I was fortunate in receiving a generous scholarship to Amherst College, where I did an honors degree in biology and fell madly in love with the elegance and precision of genetics. I declined my acceptance to medical school in order to pursue a degree in genetics at the University of Chicago. After spending a year as a postdoctoral fellow at the famous Biology Division of Oak Ridge National Laboratory in Tennessee, I decided to leave the racism of the southern U.S. to return to Canada.

My first academic position was as an assistant professor in genetics at the University of Alberta. As the most junior faculty member, I was assigned to teach genetics to second-year agriculture students. To my delight and surprise, they were the most enjoyable class I ever taught in my academic career. They asked questions about plant and animal breeding, about cloning and the future of genetic engineering—questions I hadn't studied or thought much about. So I had to do a lot of reading. When I took a position in the Department of Zoology at the University of British Columbia a year later, most of my students were premedical students. Again they asked questions that I was not prepared for—this time about hereditary disease and medical genetics—and I had to read up. And it was in the reading to answer student questions that I encountered the grotesque intersection between two great passions of my life: civil liberties and genetics.

The Dark History of Genetics

I discovered that the genetics I had been taught had been expunged of much of its history. I found that early in the twentieth century, biologists were justifiably enthralled with discoveries of principles of inheritance and their broad applicability from plants to insects to mammals. There was a sense that with these laws of heredity, scientists were acquiring the capacity to control evolution and shape the biology of organisms, including humans, at will.

Extrapolating from studies of petal color in flowers, wing shape in fruit flies and fur patterns in guinea pigs, geneticists began to make pronouncements about the role of genes in human heredity and behavior. A new discipline was created—eugenics, the science of human heredity. Eugenics was supported by leading scientists of the time and taught as a discipline in universities. There were eugenics journals and textbooks and eugenics societies. At last, it was believed, here was a solid basis on which human evolution could be directed. Positive eugenics was the increase of desirable genes in a population; negative eugenics was the decrease in incidence of undesirable genes. Not surprisingly, those traits deemed desirable were disproportionately exhibited by upper-middle-class Caucasians, whereas those that were undesirable were expressed in blacks, the poor, and criminals.

Characteristics for which hereditary claims were made included syphilis, tuberculosis, drunkenness, indolence, criminality, and deceit. Eminent scientists backed eugenics and gave it legitimacy. For example, Edward East, a Harvard professor and president of the Genetics Society of America, states in his eugenics text: "In reality, the Negro is inferior to the White. This is not hypothesis or supposition; it is a crude statement of actual fact." The problem, of course, is that "inferior" is not a scientifically meaningful category. Like "superior," "better," and "worse," it is a value judgment. In their enthusiasm and zeal for the exciting discoveries in genetics, scientists like East confused their own personal values and beliefs with scientifically demonstrated "fact."

Racism Justified by Science

With such enthusiasm and grand claims being expressed by scientists, it is not surprising that politicians noticed and began to use these ideas to justify their own prejudices. A.W. Neill, a British Columbia member of Parliament, stated in 1937: "To cross an individual of the white race with an individual of a yellow race, is to produce in nine cases out of ten, a mongrel wastrel with the worst qualities of both races." Although it was not quite a Mendelian ratio, Neill actually put a number to his claims.

In February 1941, Neill told the prime minister: "We in British Columbia are firmly convinced that once a Jap, always a Jap." In other words, it didn't matter that there were second- and third-generation Canadians who had been born and grown up in Canada. If they were Japanese genetically, Neill and others in British Columbia believed all the traits of treachery, untrustworthiness, and so on were genetically encoded. This attitude was reflected by General John deWitt, who was charged with the evacuation of Japanese Americans during World War II and said in February 1942:

> Racial affinities are not severed by migration. The Japanese race is an enemy race and while many second and third generation Japanese born on United States soil, possessed of American citizenship, have become "Americanized", the racial strains are undiluted ... It therefore follows that along the vital Pacific Coast over 112,000 potential enemies, of Japanese extraction, are at large today.

Our evacuation was justified by the claims of the scientific community about the significance of their discoveries. To my horror, I discovered as well that genetics had flourished in Germany before the war. It was scientists who helped shape some of the "progressive" legislation of the Nazi government, including the Race Purification Laws that resulted in the Holocaust. The infamous Josef Mengele was a human geneticist who held two peer-reviewed grants to carry out his study of twins at Auschwitz. By the end of World War II, revulsion at the revelation of Nazi death camps shifted the predominant opinion of geneticists to the notion that human intelligence and behavior were shaped primarily by the environment (nurture) rather than heredity (nature). The important point to remember is that this shift occurred with no significant new insight or breakthrough in science.

Biological Determinism Again

The belief in the overriding influence of nurture held until 1969, when Arthur Jensen, an educational psychologist at Berkeley, published "How Much Can We Boost IQ and Scholastic Achievement?" in the *Harvard Educational Review.* This massive study pulled together many reports of differences in IQ test scores between black and white populations. To geneticists, the question of what an IQ test actually measures is not an issue. The distribution of scores in both populations forms the familiar bell-shaped curve, but repeatedly, the two curves have a mean difference of about one standard deviation, the white means always being higher than the black means. Using extensive mathematical analyses, Jensen purported to show that the difference in means was determined primarily by heredity. His study was immediately cited by politicians such as Mississippi governor George Wallace and President Richard Nixon to justify cutting back programs such as Operation Headstart that were designed to help disadvantaged children.

Nature versus Nurture in a Racist Society

Jensen was not a geneticist. Oxford's Walter Bodmer and Stanford's Luca Cavalli-Sforza are respected population geneticists and wrote the most definitive treatise on the issue of race and IQ. If we use a less emotive illustration than race and IQ, say, the height to which bean plants grow, it can be demonstrated that there is a hereditary component in the familiar bell-shaped curve of the distribution of plant height grown from a population of seeds. That is, seeds taken from short plants will on average grow up on the short side, seeds from tall plants will be taller, and seeds from the center will fall in between. So there is a strong component of heritability determining plant height.

If we then plant one handful of seeds in moist, fertile soil and plant another handful in sandy, dry soil, we will get different results. There will be a bell-shaped distribution from both plots, and in both populations, we can show that seeds of short ones will grow up short, seeds of tall ones will be taller, and those in the middle will be in between. So in both populations, there is a strong factor of heritability in plant height. But it is absolutely incorrect to conclude that the difference in mean height between the populations reflects a component of heritability, because the only difference between them is the environment in which they are grown.

Bodmer and Cavalli-Sforza conclude that only when society is completely color-blind—when being black or white makes no difference in the way children are seen or treated—can we even begin to try to compare IQ scores to determine the component of heritability. Despite that definitive conclusion, a number of scientists, none of them geneticists, began to suggest that human intelligence and behavior have a large genetic component. Psychologist Hans Eysenck and Nobel Prize–winning biochemist Hans Krebs began to publish papers purporting to show that criminality has a high component of heritability. Harvard's Richard Herrnstein claimed that social class in a meritocracy like the U.S. is far more likely to reflect genes than work habits. And in Canada, a president of the Canadian Medical Association suggested that before receiving a welfare check, people should be sterilized to keep their genes from being perpetuated in future generations.

Molecular Genetics—History Repeats Itself

> *Where did the doctor's great project go wrong? Not in his intentions which were beneficent, but in the dangerous haste and egotistic myopia with which he pursued his goal. It is both a beautiful and terrible aspect of our humanity, this capacity to be carried away by an idea. For all the best reasons, Victor Frankenstein wished to create a new and better human type. What he knew was the secret of the creature's physical assemblage; he knew how to manipulate the material parts of nature to achieve an astonishing result. What he did not know was the secret of personality in nature. Yet he raced ahead, eager to play God, without knowing God's most divine mystery.*
>
> —Theodore Roszak

Just as eugenicists early in the twentieth century became intoxicated with the discoveries being made, molecular biologists have created a climate of belief in the basic role of genes in just about every human trait. Powerful tools to isolate and manipulate DNA confer truly revolutionary powers. Almost weekly, headlines proclaim the latest isolation of a gene for a trait, from risk taking to depression, shyness, alcoholism, and homosexuality. Few note that in the months after the initial claims, follow-ups generally fail to corroborate the original claim or they show that hereditary involvement is far more complex.

One of the big mistakes made is a confusion between correlation and causation. Take, for example, the gene controlling the enzyme alcohol dehydrogenase (adh). There exist two different states of this gene, adh^A and adh^B. Suppose a study demonstrates that 80 percent of alcoholics have the adh^A gene whereas 80 percent of nonalcoholics have adh^B. That is a *correlation*. But it is completely incorrect to conclude that adh^A *causes* alcoholism, yet the press and even scientists themselves frequently fall into that trap.

Think of it this way. Suppose you study all people who die from lung cancer in Vancouver over the past ten years and discover that 90 percent of them had stained, yellow fingers and teeth. That is a correlation. Who would ever conclude that stained yellow fingers and teeth *cause* lung cancer? Yet it happens over and over when molecular biologists isolate fragments of DNA that correlate with traits, some as diverse and complex in expression as homosexuality.

Rapid Growth of Revolutionary Science

Genetic engineering is a truly revolutionary area of science, made possible by the incredible speed and power of newly acquired techniques. When my daughter was in her last undergraduate year of university, she isolated, sequenced, and compared mitochondrial DNA of three geographically separated but related plant species for a senior research project! It was breathtaking to me because such experiments were inconceivable when I graduated forty years earlier. So I understand why there is so much excitement. I too am excited and have followed genetic engineering vicariously for many years. But in a revolutionary area where excitement abounds, history informs us there is all the more reason to encourage vigorous debate and to be critical and cautious.

By the 1970s, it had become clear to me that molecular genetics was going to revolutionize the field and have profound social ramifications. *Biotechnology* refers to the field of applied genetics wherein molecular manipulations are carried out in living cells and organisms. The impetus for biotechnology was the ability to make combinations of DNA molecules from diverse species and to test those molecules in living cells. The technique was known as recombinant DNA. As a columnist for the National Research Council of Canada publication *Science Forum,* I wrote in 1977:

> For young scientists who are under enormous pressure to publish to secure a faculty position, tenure or promotion, and for established scientists with "Nobelitis," the siren's call of recombinant DNA is irresistible.... In my own laboratory, there is now considerable pressure to clone *Drosophila* DNA sequences in *E. coli*.... My students and postdocs take experiments and techniques for granted that were undreamed of five or ten years ago. We feel that we're on the verge of really understanding the arrangement, structure and regulation of genes in chromosomes. In this climate of enthusiasm and excitement, scientists are finding the debate over regulation and longterm implications of recombinant DNA a frustrating roadblock to getting on with the research.

A year later, having encountered little support within the scientific community to engage in critical discussion about the social, moral, and ethical implications of recombinant DNA, I tried to explain the reluctance in *Science Forum:*

> I can appreciate the pressures that are brought to bear to stifle dissent within the scientific community. Peer approval brings with it invitations to give lectures, to speak at symposia and honorary positions in scientific organizations. The driving priorities of young scientists are to get and keep good sized grants and achieve recognition, tenure and promotion. Therefore, outspoken criticism is understandably rare in this group and they depend on people higher up in the scientific hierarchy to set their objectives.... What am I

> saying? Not that scientists are evil, malicious or irresponsible—rather that our personal priorities, membership in a vested interest group, ambitions and goals prevent us from objectively weighing the social against personal consequences of our work.

A Personal Moratorium on Genetic Engineering

My own personal experience with the consequences of well-intentioned but scientifically unjustified claims and the insights I had gained while trying to answer my students' questions had made it clear to me that there was a very important need for scientists to engage in public discussions about the significance and implications of their work. Because I wanted to be able to participate credibly in this discussion, I declared in a 1977 column in *Science Forum:*

> Can the important questions be addressed objectively when one has such high stakes in continuing the work? I doubt it. Therefore, I feel compelled to take the position that ... no such experiments [on recombinant DNA] will be done in my lab; reports of such experiments will not acknowledge support by money from my grants; and I will not knowingly be listed as an author of a paper involving recombinant DNA.

I had achieved far more in science than I had ever dreamed of or hoped for. It had been the joy of research that absorbed me for a quarter of a century. I loved the excitement and camaraderie of the lab. I was proud of our group, at one time the largest in Canada, and the work we did.

But it was the muddy area of extrapolation of scientific insights to broader society that concerned me, because there were numerous examples of individuals making claims far beyond their scientific legitimacy. I felt that some of us whose careers and reputations were not in jeopardy had to forgo this work in order to take part, as scientists, in the discussions of the moral and ethical questions free from the bias of vested interest in the work. Scientists working for the nuclear, tobacco, and petrochemical industries, either as employees or recipients of research grants, speak from a perspective of those with a stake in continuing income and research support, and therefore it's natural that they would tend to deflect criticism rather than discuss it openly. There was no reason to suppose that scientists in biotechnology would be any different. Eventually I stopped taking government grants altogether because grants are awarded by peers, almost all of whom are promoters of research without regard to social or ethical concerns. I didn't want to be dependent on, and thus vulnerable to, the influence of outside agendas.

Damned If I Do, Damned If I Don't

As a popularizer of science through newspaper columns, television, and radio, I am able to survey a far broader range of topics and questions than I ever did as a research scientist. Rather than losing my interest in the field of biotechnology and all of its implications, I have a broader perspective to reflect upon and have written extensively on the subject over the years. In 1986, I discussed moral and ethical issues of genetics in my autobiography, *Metamorphosis: Stages in a Life.* I wrote syndicated columns on genetics that became chapters in the best-selling books *Inventing the Future* and *Time to Change.* In 1988, science writer Peter Knudtson and I coined the term *genethics* and co-wrote *Genethics: The Ethics of Engineering Life,* which became a best-seller and continues to be widely used in university courses.

So it has been puzzling to me when individuals, some not even scientists, but spokespeople for the biotechnology industry, call my credibility into question. I deliberately gave up the day-to-day excitement of scientific research to remain a credible discussant on the moral and ethical implications of the new genetics. But I didn't forget all I'd learned and practiced as a scientist. At the very least, all of us who participate in the discussion ought to be forthright about the sources of our funding, our position in companies, and any other factors that might influence our perspectives and bias our statements.

Biotechnology Is Here

Today products of biotechnology are being rammed into our food, onto our fields, and into our medicines, without any public participation in discussions and with the complicity—indeed, the active support and funding—of governments. But there are profound health, ecological, and economic ramifications of this activity. At the heart of biotechnology is the ability to manipulate the very blueprint of life, removing and inserting segments into diverse species for specified ends. While plant and animal breeding over the past ten millennia has built the agriculture we depend on, biotechnology takes us far beyond the crude techniques of breed and select. It behooves us, therefore, to examine the underpinnings of the claims, the potential, and the limits of this young field.

Biotechnology to Feed the World

Perhaps the most frequently cited rationale to get on with genetic engineering as rapidly as possible goes like this: Human population continues to increase by more than 80 million a year, most in the developing world. To avoid clearing more forests and draining wetlands to meet the needs of this burgeoning population, proponents argue, the only option to protect nature and feed the masses is to increase yields per hectare through biotechnology.

This argument carries a lot of weight, despite the irony that the number of people suffering from severe malnutrition is about equal to the number of people afflicted with obesity in the rich nations. However, biotechnology is being driven by vast sums of speculative money. To justify those investments and to attract even more money, a product is needed. That's why so many companies have already foundered—they've failed to live up to the expectations of a product. The very survival of biotech companies depends on the expectation of profits from the company's products, whether pharmaceutical or agricultural. Those products are made at enormous cost. In the case of food biotech, the people who are most desperately in need of food are also the poorest. James Wolfensohn, president of the World Bank, claims that 1.3 billion people exist on a dollar or less a day, while 3 billion struggle on two dollars or less daily. It would be a breathtaking reversal if free-enterprise capitalists were suddenly overwhelmed with generosity and concern for those less well off and made genetically engineered products available at prices the needy can afford. Feeding the starving masses through biotech is a cruel hoax that cannot be taken seriously.

The Real Nature of Scientific Knowledge

I have no doubt that important products will come out of genetic engineering—but in the more distant future. It is the profit-driven rush to grow genetically engineered organisms in fields, where they might contaminate other species, and to introduce new products into the market that is most disturbing. My major concerns are based on simple principles. Every scientist should understand that in any young, revolutionary discipline, most of the current ideas in the area are tentative and will fail to stand up to scrutiny over time. In other words, the bulk of the latest notions are wrong. This is by no means a knock on science; it is simply an acknowledgment that science progresses by demonstrating that current ideas are wrong or off the mark. The rush to exploit new products will be based on inaccurate hypotheses, so supposed benefits are questionable and could be downright dangerous.

I graduated as a fully licensed geneticist (that is, I had a Ph.D.) in 1961. It was eight years after Watson and Crick's famous paper, and we had learned a lot—we knew about DNA, the number of human chromosomes, the operon, and so on. But today, when I tell undergraduates about the hottest ideas of chromosome structure and gene regulation in 1961, they laugh in disbelief. In 2003, the best notions of forty years earlier seem naïve and far from the mark. But those students are less amused when I suggest that twenty years from now, when they are established scientists, the ideas they are excited by now will seem every bit as quaint as the ones I was excited by in my early days. In any new area, scientists make a series of observations and then set up a hypothesis that makes sense of the observations. That hypothesis enables a researcher to design experiments to test its validity. When the experiment is performed and the data gathered, chances are that the hypothesis will be discarded or radically altered, and then further experiments will be suggested. That's how science proceeds. But that procedure suggests we ought to proceed with far less haste. This sentiment is reinforced by Roger Perlmutter, executive vice-president of research and development for the biotech company Amgen: "Things we take as the absolute truth now are going to look pretty silly a few years from now." He's right on, but then the question is, why rush to exploit ideas that will turn out to be "pretty silly"? Isn't that foolhardy, even dangerous?

Not Quite Ready

When a biotechnologist can clip out or synthesize a specific sequence of DNA, insert it at a precisely specified position in a host genome, and obtain the predicted expression of the inserted DNA with no other complications, then we can say that it is a "mature" discipline. But when that happens, one can't publish papers on such a manipulation because it will be old hat. If you've checked biotech publications these days, you'll be amazed at their number and variety. Those reports are based on experiments in which the researchers *didn't know* what the results would be; after all, that's why experiments are done and reported. Doesn't the abundance of biotech papers inform us that we still have a huge amount to learn? That suggests strongly that the discipline is far from mature enough to leave the lab or find a niche in the market.

The problem with biotechnology as it's presented today is that those pushing its benefits stand to gain enormously from it. I believe that they start from a sincere faith in the benefits and in our ability to "manage" the genetically engineered organisms and products safely. But we've learned from experience with the tobacco, nuclear, petrochemical, automobile, and pharmaceutical industries and military establishments that vested interest shapes a spokesperson's perspective and precludes an ability to examine criticisms or concerns in an open fashion.

Linear Science—An Illusion

Promoters of biotechnology foster a version of how science proceeds that is totally at odds with real science. They confuse the way scientists write grant requests with reality. The game scientists play in grant applications is to act as if the money will be used to do experiment A, which will lead to experiment

B and on to C and D, and then, voilà, a cure for cancer. Scientists perpetuate the illusion that science progresses this way as justification for receiving a grant. It's as if scientific discovery proceeds in a linear way—but nothing could be further from the truth. Experiment A is carried out because the researcher doesn't know what the results will be and so has no idea where the results and then subsequent experiments will lead. That's why despite all of the hoopla over biotechnology, so few concrete products have come forth and there is considerable controversy surrounding those that have reached test plots or the marketplace.

The great strength of science is in *description*. We discover things wherever we look because despite the enormous growth of science in the twentieth century, our knowledge of how the world around us works is still minuscule.

DDT—A Case Study

The fatal weakness of science is in *prescription* of solutions. A classic example is DDT, a complex ring molecule first synthesized in the nineteenth century. In the 1930s, Paul Mueller found that it kills insects. The power of chemistry to control a scourge that had plagued humankind since the beginning of time was widely trumpeted. At the time Mueller made his discovery, geneticists could have suggested that using an insecticide would simply select insects carrying mutations conferring resistance to the chemical. They would quickly replace the sensitive strains and thereby set farmers on a treadmill of requiring an endless string of different pesticides. Ecologists of that time could also have suggested that of all animals in the world, insects are the most numerous and diverse and play critical ecological roles such as pollination, predation, and feeding of other species. Perhaps one or two insect species per thousand species are pests to human beings. Using a broad-spectrum insecticide to get at the one or two species that are a nuisance to humans seems analogous to killing everyone in a city to control crime—pretty crude and unacceptable.

But in their exuberance over the power of chemistry, geneticists and ecologists failed to raise these concerns, millions of kilograms of DDT were manufactured and used, and Paul Mueller won a Nobel Prize in 1948. Years later, biologists discovered biomagnification of DDT up the food chain that eventually affects fish, birds, and mammals.

The history of DDT and later, CFCs, reveals that we are very clever at applying scientific insights for specific purposes, but the repercussions in the real world (for example, biomagnification and ozone depletion) could not be predicted beforehand and were only discovered after widespread use. There is absolutely no reason to think genetically engineered organisms and products will be free of such unexpected consequences.

A Clockwork Universe

Ever since Isaac Newton and René Descartes, scientists have assumed the cosmos is like an immense mechanical construct whose components can be examined piece by piece. If this is so, then, in principle, we can learn about parts of nature and eventually acquire enough knowledge of the fragments that we could put them all together to recover a picture of the whole. Biologists have been especially critical of any suggestion that the whole is greater than the sum of its parts; they see this idea as an expression of vitalism, a discredited notion that living organisms possess a kind of vital essence absent in nonlife. Few biologists are concerned with the problem that life arose from the aggregation of nonliving matter, even though the state of aliveness cannot be anticipated from the properties of the nonliving components.

Reductionism, the focusing on parts with the goal of understanding the whole of a mechanistic universe, has been a productive methodological approach. Thus, scientists focus on a subatomic particle, an atom, a gene, a cell, or tissue, separate it from everything else, control everything impinging on that fragment, measure everything within it, and thereby acquire profound insights—into that fragment. But physicists learned early in the last century that parts interact synergistically, so new properties emerge when you combine them that could not be anticipated from their individual properties. After defining all of the physical properties of atomic hydrogen and atomic oxygen, physicists would be at a total loss to anticipate the properties when two atoms of hydrogen are combined with one atom of oxygen to make a molecule of water. Biologists and doctors have yet to internalize that understanding. Thus, it was long assumed that by studying a chimpanzee in a cage, for example, one could learn everything there was to know about the species. It was only when Jane Goodall went out into the field and studied chimps in their natural habitat that she discovered a completely different animal. Biophysicist Brian Goodwin has shown that the collective behavior of ants within a colony cannot be explained by the sum of the behavior of individuals of each caste.

Missing the Whole by Focusing on Parts

In focusing, we lose sight of the rhythms, patterns, cycles, and context that make the object of study interesting in the first place. Biotechnology is the ultimate expression of reductionism, the faith that the behavior of individual pieces of DNA can be anticipated by studying them individually. Richard Strohman, a leading scientist and former chair of the Department of Molecular and Cell Biology at Berkeley, stated the problem this way:

> When you insert a single gene into a plant or an animal, the technology will work ... you'll get the desired characteristic. But you will also ... have produced changes in the cell or the organism as a whole that are unpredictable.... Genes exist in networks, interactive networks which have a logic of their own.... And the fact that the industry folks don't deal with these networks is what makes their science incomplete and dangerous.... We are in a crisis position where we know the weakness of the genetic concept, but we don't know how to incorporate it into a new, more complete understanding.

Biotechnologists assume all pieces of DNA can be removed and inserted as if they are equivalent. But as Strohman points out, genes don't exist as independent entities, they exist within complex sets of networks. From the moment of fertilization, whole suites of genes are turned on and off in an orchestrated sequence that leads to the development and differentiation of an individual. It is ultimately the total expression of that sequence and suite of genes that produces the phenotype—the visible characteristics—of the organism, and that is what natural selection acts upon. So the genome should not be seen as a bunch of individually functioning and selected genes; they act in concert. Biotechnologists assume that they can simply take a gene from a flounder, for example, and stick it into a tomato plant, where it will function and produce a predictable result. But that strikes me as comparable to taking Bono out of U2, sticking him into the New York Philharmonic Orchestra, and asking him to play his music while the other musicians play theirs. They will all be playing music, but how it will all sound together cannot be anticipated.

Craig Venter is a brash entrepreneur who sought to finish decoding the sequence of the three billion letters of the human genome before anyone else. Setting up his own company, Celera, he created tremendous controversy and stimulated an acceleration of the project. Yet even such a booster of biotechnology as he admitted in 2000: "We know far less than 1% of what will be known about biology, human physiology and medicine. My view of biology is 'we don't know shit.'" Venter amplified the implications of his remark the following year:

> In everyday language, the talk is about a gene for this and a gene for that. We are now finding that this is rarely so. The number of genes that can work in that way can almost be counted on your fingers, because we are just not hardwired that way. You cannot define the function of genes without defining the influence of the environment. The notion that one gene equals one disease or that one gene produces one key protein, is flying out the window.

With so much yet to be learned, the rush to exploit biotechnology can only be seen as wrongheaded.

Unknowing Participants in an Experiment

The growth of genetically engineered plants over vast areas of the Prairies is already a fait accompli. Pressured by companies like Monsanto, the Canadian and U.S. governments have acted as cheerleaders for the biotech industry, approving new strains with little regard to the urgent questions that have been raised. Unlike chemical pollutants or radioisotopes, which degrade or decay, genetically engineered plants and animals reproduce and mutate. Once they are released into nature, they cannot be recalled.

The impressive feature of life on Earth is its tenacity. Despite all the changes—the Sun is 30 percent warmer now than it was 4 billion years ago, ice ages have come and gone, continents have collided and generated mountains and oceans, magnetic poles have reversed and rereversed—life has persisted and flourished over 3.8 billion years. Once it has a hold, life is incredibly tenacious. Wind, insects, rains, rivers—many factors can act as a vehicle for genetically engineered organisms to spread their genes. Lavern Affleck, a Saskatchewan farmer, testified before a New Zealand committee examining the benefits and hazards of genetically engineered crops:

> Canada has gone blindly into broad scale experimentation with the Canadian land base. It is an experiment which cannot be retracted, and was entered into without sincere reflection as to possible ramifications. In our experience, crops (and weeds) are spread in so many ways (wind, the waterways, on the roadside, on farm machinery and trucks) that it is impossible to prevent accidental releases into unwanted areas. We now have some degree of GE [genetically engineered] crop contamination across our entire Canadian prairie land base.

Biotechnology and Society

Lacking an understanding of the complex relationship between scientific research and its application, governments sporadically commit money to specific areas in the hopes of stimulating economic benefits. (In my opinion, they are doomed to failure for a number of reasons; but that is not the point of this essay.) But now, federal and provincial and state governments have latched onto the life sciences, promoting biotech companies and their products. Unfortunately, molecular biology is an arcane discipline that few nonspecialists can decipher. Biotechnology companies and scientists doing molecular research are aggressively proclaiming the benefits of their work. In their zeal, objections and concerns are brushed aside as trivial or baseless, just as the tobacco industry dismissed health concerns about smoking. But how can society deal with new discoveries and applications in ways that will minimize hazards to people and ecosystems?

In my view, universities are places where these issues should be openly discussed and debated. University scientists straddle the scientific disciplines, speaking the arcane language of science while communicating with students and the larger society in nonjargon vernacular. A university is a very special institution in society—a community of scholars and students exploring ideas at the very cutting edge of human thought. Many of these ideas are perceived as dangerous to society, and thinkers are often viewed as threats to the established order. To ensure the freedom for scholars to pursue their work and protect them from outside interference, universities confer the privilege of tenure. Tenure brings with it the responsibility to share knowledge and speak out on issues where a scholar's field impinges on society.

Sadly, universities have compromised this position by entering into extensive partnerships with the private sector. In their search for funds, university administrators have found sources in corporations and now actively encourage faculty to establish companies that will provide royalties to the university. The deleterious consequences can be seen at the University of British Columbia Faculty of Forestry, where the foyer of the building is filled with plaques acknowledging the contribution of forest companies. While environmentalists have for decades decried British Columbia's clearcut logging practices as both destructive and unscientific, UBC's forestry faculty has largely toed the industry line. Supporting the forest industry seems to have become more important to the forestry faculty than reasoned debate. (Interestingly, the recent growth in numbers of women in the faculty has been accompanied by more genuine interest in and debate about alternative forestry practices.)

The same promotion of industry perspectives occurs in those faculties receiving money from pharmaceutical, chemical, and military sources. Students are presented with one-sided propaganda about the potential benefits of these areas with little balance from those with concerns. Indeed, most faculty members who do have reservations seldom dare to speak out, or if they have the courage to risk the approbation of their peers, they suffer all manner of indignities. In my experience, merely questioning the activity or suggesting possible hazards is to invite strong disapproval and accusation of being "antiscience" or "emotional and nonscientific." It is a sad state in a community of so-called scholars, where dissent or difference of opinions is supposed to be valued. One way to raise important issues without being overwhelmed by pro-biotech lobbying would be through a Royal Commission or Congressional inquiry to examine the broad societal, health, ecological, and economic implications of genetic engineering.

In Europe, where a "slow food" movement has sprung up as a counter to North American fast food, genetically engineered crops and food have been kept out of the continent. Europeans have applied the Precautionary Principle, which demands convincing evidence of both a need for the product and its safety before acceptance of a new technology. Europeans tell me they are watching North Americans for evidence of hazard or safety because we "are doing the experiment." Canadians have been eating genetically engineered food for more than five years without being informed or provided information on labels.

Over recent decades, it has been revealed that until the 1960s scientists carried out experiments on unwitting human subjects. A few examples: patients infected with syphilis were deliberately denied treatment in order to follow the full course of the disease; inmates of mental asylums were administered the hallucinogen LSD to determine the effects; people judged mentally or physically handicapped for genetic reasons were sterilized. Out of these examples of excessive scientific exuberance, scientists have accepted the conditions for carrying out tests with humans: prospective subjects must first be fully informed of what is to be done, and the subjects must give approval before the study is carried out. Convinced by the biotech industry that genetically engineered foods are "substantially equivalent" to non–genetically engineered food, governments have demanded little in large-scale studies of the long-term effects of ingesting such food. (The one experiment by Dr. Arpad Pusztai that showed deleterious consequences of feeding genetically engineered potatoes to rats was peer-reviewed and published in the prestigious medical journal *Lancet*. Discounted by the media and industry lobbyists, Pusztai's experiments remain the only feeding study with genetically engineered foods published.) So, in Canada and the U.S., large numbers of people are being subjected to a massive experiment without providing informed consent. At the very least, all people should be able to see on labels what food is genetically engineered so that they can make their own choice.

A Future for Biotechnology?

As a geneticist, I continue to take enormous vicarious delight in the incredible technological dexterity being gained and the acquisition of answers to basic biological questions I never thought I would live to see solved. The exuberance of geneticists is understandable, especially when there seem to be such opportunities to engineer life according to our specifications. I have no doubt there will be important uses of these techniques and insights in the future. But we still have an enormous amount to learn. Already there are reports of and experiences with genetically engineered crops in open fields and in our food that suggest we have valid reasons to proceed with greater caution.

It behooves every scientist to remember the experience of the nuclear industry. During World War II, allied scientists rushed to build an atomic bomb before the enemy succeeded in building its own. Once the bomb was built, the Allies learned the enemy was not in the race. Atomic bombs represented a radically new weapon that not only increased the scale of destruction on that generation but afflicted future generations with a legacy of induced genetic alterations. Nevertheless, the use of these revolutionary weapons was justified by the potential to save lives by completing the war more quickly. Years *after* atomic bombs were created and used, scientists discovered new phenomena: radioactive fallout and bioaccumulation of radioisotopes, electromagnetic pulses of gamma rays that incapacitate electrical connections, and the vast ecological consequences of nuclear winter. There is absolutely no reason to suppose that biologists know enough to anticipate the ecological and health ramifications of a revolutionary technology such as genetic engineering. Governments must resist the economic pressures and show leadership and concern for the long-term health of people and nature. And scientists involved in this exciting area should learn from history and welcome free and open discussion about ecological, health, and social implications of their work.

It Always Costs

I HAVE LONG BELIEVED THAT WE HAVE TO HAVE GREATER SCIENTIFIC literacy at all levels of society if we are to have any hope of controlling the way science and technology affect our lives. That's why I went into broadcasting.

But I have only recently realized that my underlying faith in the power of greater awareness is misplaced. First we must understand that there is no such thing as a problem-free technology. However beneficent, technology always has a cost.

Think, for example, of DDT—it killed malaria-carrying mosquitoes in huge numbers and without question saved millions of lives in tropical countries. But geneticists could have predicted that DDT would exert incredible selective pressure for mutations that would confer resistance to DDT on the mosquitoes and that within a few years large numbers would return. They did. But once committed to a chemical approach, we had to turn to other, more toxic compounds.

The ecological damage from massive use of chemical sprays has been enormous because DDT is not specific and kills all insects. Furthermore, the compound is ingested by microorganisms, which in turn are eaten by larger predators, and so on up the food chain. Thus, initially minute quantities become more concentrated up the food chain in a process called *biomagnification*. The final result was that DDT ended up in the shell glands of birds, affecting the thickness of eggshells and eventually causing heavy bird mortality.

There are numerous examples of how technological innovations have had detrimental side effects that eventually outweighed their benefits. I used to think that what we needed was some kind of vehicle, like panels of citizens representing a broad range of interests, to do a cost/benefit analysis of all new technologies. The idea was that by carefully weighing the benefits and bad side effects, we could make a more informed decision about whether to allow a new technology to be used. My belief that this would help us avoid future problems was based on faith in our predictive capabilities. Indeed, much of the testing of environmental and health impacts is made on that faith. But we can't rely on such a system.

For one thing, our assessments are always limited. For example, suppose we do an environmental impact assessment of drilling for oil in the High Arctic. The studies, of necessity, are carried out in a limited time within a restricted area. It is simply assumed that scaling up the observed effects of the two drill holes by a factor of one hundred or more gives a reasonable estimate of the impact of major exploration.

Well, some effects are called *synergistic:* several components interact to give new or greater effects than the sum of their individual impacts. Also, during an assessment, you can bet industry will be on its best behavior, so the results will always be on the conservative side.

It is also true that even if a study is made over ten years (which it won't be), we could never anticipate all the fluctuations of conditions in this sensitive area. I've known colleagues who have studied populations of animals or plants over decades and find nice cycles and patterns that are predictable until suddenly completely unexpected fluctuations occur. They get out more publications that way, but we ought, then, to be a lot more humble about how little we know.

For another thing, we know that major blowouts, spills, or accidents are relatively rare. Suppose one happens an average of once every twenty holes. By studying two holes and finding no effect, we are not justified in concluding that drilling one hundred holes will also be accident free. It would be just as invalid were an accident to happen in one of the test holes to conclude that half of all drilling sites will have a bad episode. The numbers are statistically meaningless.

Food additives, pesticides, and drugs are extensively tested before they are approved for use. But numerous cases inform us that we can't anticipate all the effects. The DDT example is classic—at the time it was used, we didn't even know about biomagnification, let alone the chemical's concentration in bird shell glands.

Remember thalidomide or DES? Or consider the oral contraceptive. It had been extensively tested (in Puerto Rico) and shown to be efficacious without detrimental side effects. It was only after millions of healthy, normal women had taken the pill for years that epidemiologists could see negative effects. No amount of pretesting could have anticipated them.

So we come to a terrible conclusion. Technology has enormous benefits. They are undeniable—that's why we're hooked on it. Once a technology is in place, it becomes impossible to do without it; we can't go back to doing things the old way. But the pretesting of any new technology is flawed because it provides only a limited assessment of its impact. The tests are restricted in size, scope, and time and are based on what we decide a priori might be a possible effect.

But how can we test for something that we don't know will happen? If every technology has a cost, the more powerful the technology, the greater its potential cost. We have to build into our judgments a large leeway for our ignorance and err on the side of extreme caution. And perhaps it's time to realize we don't have to do everything just because we can.

The Illusory Oil Change

THE ENVIRONMENTAL CONSEQUENCES OF OUR TECHNOLOGICAL SOciety, with its high consumption and disposable products, are forcing us to reassess our lifestyles. Recycling was once the rule, dictated simply by necessity. We have forgotten that our parents conserved routinely, and now we are being haunted by the consequences of our profligate ways.

We've all taken our cars in to have the oil changed—perhaps you've even changed the oil yourself. But have you ever wondered what happens to that yucky black stuff that was drained out? That lubricating oil is potentially reusable, but most of it ends up dumped on the ground or in rivers and creeks. It's a classic example of our society's shortsightedness.

We live in a strange world of illusion. The current prices of oil are depressed because of a "glut" on the market, yet every oilperson knows oil is going to run out early in the twenty-first century. We have enormous environmental problems, yet we continue to pay no attention to the destructive effects of many of our products that end up polluting. That brings us back to used oil.

There are two kinds of lubricating oil: the stuff we use in our cars and industrial oils. Of the 200 million gallons of lubricating oil produced in Canada annually, half is used up in the lubricating process, but the other 100 million gallons are potentially recoverable. In fact, about 37 million gallons are collected, of which about 22 million gallons are re-refined and the remainder is burned or spread on roads. What about the uncollected 63 million gallons? Chances are they go into sewers or onto the ground. Without a doubt we end up drinking it in our water and eating it in our vegetables and meat. So not only do we waste a precious resource by failing to recycle all this used oil, it is also a major contaminant of the environment.

Used oil is also laced with deleterious chemicals that are removed in the re-refining process but that are liberated when the used oil is burned in low-grade furnaces or dumped. Some of those contaminants include lead, zinc, chromium, arsenic, chlorine, bromine, PCBS, polycyclic aromatics, and volatile and semivolatile organics—a rather nasty gallery of chemicals.

It costs money to re-refine oil. There has to be a system whereby it can be stored, picked up, and transported to re-refining plants. There is little incentive, especially when "virgin" oil—refined crude oil—is so cheap. Twenty years ago in the U.S., there were over two hundred re-refiners. In 1987, only three remained, and they were struggling to survive. At the same time in Canada, there were six, and they were all barely making it.

Part of the problem is psychological—North Americans believe that re-refined oil is lower quality than virgin oil. Yet a study by the National Research Council of Canada showed that re-refined oil is as good as or even better than the refined. But we are reluctant to purchase the re-refined, especially since it is more expensive.

A major part of the problem for re-refiners is political—all the tax incentives and subsidies go to the discovery and exploitation of crude oil. There are no economic incentives for the re-refiners. There should be every encouragement to conserve through recycling and to protect the environment by removing toxic contaminants and preventing the introduction of the oil into the environment. But therein lies the problem. The producer of a product—any product—usually has no obligation to anticipate its *total* cost, including eventual disposal, yet that should be built into the initial costing.

The best example of our myopia is the nuclear industry, which built plants long before there was any serious consideration of disposal of radioactive wastes or decommissioning aging plants. Economic or legal incentives to recycle in an environmentally responsible way are needed. Re-refining oil ought to be an apple pie issue, but the industry is on the ropes.

We have to get over the idea that we can dump liquid waste into sewers and forget about it or that we can slop it onto the ground where it will be absorbed. It is ironic that while the PCB spill near Kenora, Ontario, created a public outcry, *millions* of gallons of used oil—much of it containing PCBS and other toxic substances—are sprayed onto dirt roads to keep down the dust. More goes into our waters, where it is estimated that the cost is up to $8 million a year to repair corrosion and replace filtration systems. Millions of gallons of used oil are also burned in furnaces for low-grade heating of greenhouses. The burning temperature is too low to destroy PCBS, which are simply liberated into the air and often end up being absorbed by the plants growing in the greenhouses. Some used oil is actually sprayed onto pigs' backs to keep them from getting sunburned in the summer.

Whatever we take from the ecosystem denies it to other life-forms, and whatever we put into it flows through the various cycles of air, water, and soil. For millennia, our numbers were small and our technology simple, so the environment seemed limitless and endlessly self-cleansing. Today we are too numerous and our technology is too powerful for nature to be as forgiving. Governments don't offer enough rewards to those who conserve resources for future generations or adequately punish those who use up or damage the environment. It doesn't make sense to recycle only if it is economically profitable: we live on a finite planet where all life is interconnected.

So the next time you empty your crankcase, think hard before dumping that stuff down the drain.

Nuclear Menus (Or, Eating in the Nuclear Age)

HOW MANY TIMES HAVE YOU DISCOVERED AN OLD BAG OF ONIONS OR potatoes that are no longer edible because they've sprouted roots or leaves? Ever tried to keep meat from spoiling without refrigeration? Throughout history, food spoilage has been a constant concern, and we've come up with ingenious solutions—smoking, salting, dehydration, bottling, refrigeration, freezing, and so on. And now we have the latest, a miracle of twentieth-century technology, irradiation.

Decay is brought on by microorganisms that parasitize dead plant or animal tissue. Dark, humid, and warm storage conditions may stimulate growth of vegetables. Both of these processes—decay and sprouting—can be inhibited by massive doses of radiation that kill decay-causing organisms or knock out a plant's regenerative capacity. Food irradiation is being touted as a technological revolution that could feed the world's hungry and, not incidentally, rescue an ailing nuclear industry.

The liberation of energy by splitting atoms had been the crowning achievement of a hitherto esoteric science, and after the war, physicists hoped for an era of cheap, clean, limitless energy by using "atoms for peace." Atomic Energy of Canada Limited (AECL), a Crown corporation, has spearheaded the peaceful use of nuclear energy. Canada developed and is the major producer of Cobalt 60 "bombs" for fighting cancer. And our Candu reactor is claimed to be the best in the world. But Candu has become an economic black hole: we have poured billions of tax dollars into this technology with no hope of ever recovering the money. A dependable consequence of megaprojects is that once started they develop a life of their own and become difficult to shut off.

Nuclear technology is complex and represents the ultimate practical achievement by science and engineering. But Three Mile Island and Chernobyl put to rest for good the faith that any technology can be "foolproof." Human beings will always outfool the cleverest fail-safe devices. The nuclear industry has been beleaguered by public concern about safety, economics, and nuclear weapons. So it's not surprising that the nuclear establishment is looking for new justifications and potential revenue. And that brings us back to food irradiation. AECL and the international nuclear community have been hard at work pressing the case for the economic and health benefits of food irradiation. The Science Council of Canada and leading scientists have strongly supported the claims of safety and the benefits of food irradiation. In the U.S., the Centers for Disease Control (CDC), the Assistant Secretary of Health, the Department of Agriculture, and the Food and Drug Administration (FDA) have all endorsed the safety of food irradiation.

Now I have to be honest. I've been neutral about food irradiation because I have looked at scientific papers on fruit flies fed irradiated food and have not been impressed by data purporting to show that such food is mutagenic. But it's time to get off the fence.

One reason is my inclination to mistrust any grand claims being made by people with a strong vested interest in whatever they're pushing. Few of us are convinced any longer by the tobacco industry's claims about the benefits and safety of its products. Shouldn't we be equally cautious about the zealous sales job being done by people in the nuclear industry? Don't get me wrong—I don't think they are evil or dishonest. But they are *believers,* and believers often use every means, including exaggeration, distortion, and intimidation, to sell the product they have faith in. We must take the nuclear industry's denial of possible health hazards of irradiated food with skepticism and look to other sources of information.

These days we are flooded with drugs, pollutants, and additives in our food, air, and water. We should know as much as possible about what we ingest. Why, then, is the food irradiation industry so anxious to hide its activity by deleting any reference to irradiation on food labels? It is insupportable to deny people the right to be informed and make up their own minds. If people mistakenly assume irradiated food is radioactive, then it's up to the industry to educate them, not deliberately keep them uninformed.

History teaches that we simply cannot predict the ultimate costs and benefits of new technologies. If any industry should know that, it's the nuclear industry. The long-term consequences of nuclear explosions (fallout, holes in the ozone layer, electromagnetic pulses, nuclear winter) were only discovered long after the weapons had been used. The entire history of technology is full of examples in which immediate benefits were obvious but the costs completely hidden and unpredictable a priori. If we can't anticipate what the effects will be, we don't even know what to look for. I think as long as we get caught up in trying to marshal evidence from animal studies, epidemiology, and other areas to show harmful effects, the potential consumer and eventual taxpayer are bound to lose.

The wrong questions are being asked about food irradiation. It is not a debate over whether the practice is safe—we won't know until the technology is widely used and millions of people have eaten the food for years. If history is any guide, there will be unexpected and deleterious effects. Right now a major motivation for food irradiation is to keep an ailing nuclear industry alive. That obscures the fundamental question: do we need a nuclear industry in this day and age? Surely a Royal Commission to reassess the role of nuclear energy in our society is long overdue.

The Prostitution of Academia

ALL GOVERNMENTS OF INDUSTRIALIZED COUNTRIES WISH TO EMULATE Japanese success in the high-tech industries, so they are attempting to capitalize on the creative energies of scientists in universities.

Responding to pressure from the government and industry, Canadian universities are encouraging academics to develop ties with the private sector, thereby accelerating the transfer of basic knowledge to industry. The unique role of academic scholars as a group without a vested interest in business or government is thus terribly compromised.

In a glossy advertisement for the University of British Columbia entitled "Engine of Recovery," then-president David Strangway states on the first page: "Universities are a major source of free enquiry, providing the ideas that can later be exploited by free enterprise. We need both the push of free enquiry and the pull of free enterprise for success in our society." The rest of the brochure is filled with examples of people, primarily scientists, ostensibly solving practical problems in medicine, industry, and society.

Across North America, universities are rushing to become part of the industrial enterprise, as faculty are being encouraged to become entrepreneurs who exploit their discoveries for profit. There have been few objections to or questions raised about this process. I, for one, do not agree with President Strangway's political-economic analysis of the societal role of universities, and I have grave concerns about the headlong rush to industrialize the university. Let me explain.

Historically, universities were never meant to be places where people prepared for jobs or where specialists aimed to benefit the "private sector." The university has traditionally been a community of people sharing in the exploration of human thought and creativity. The common assumption since universities became public enterprises has been that if the best minds of our youth are an important natural resource, then universities will maximize their development.

A good university is a place where scholars, dreamers, artists, and inventors can exist with no more justification than excelling at what they do and sharing their skills and knowledge. The full range of human thought is encompassed within a university. One consequence is that such knowledge often leads to criticism of government and industry. University scholars can be a pain in the neck to people in power. That's why academics have fought for tenure as a means of protection from harassment for their ideas and social critiques. Society needs objective critics if it is to have more than parochial, self-centered goals. Sadly for most North American academics, tenure has become a sinecure rather than a privilege and opportunity.

The industrialization of the university is a mistake for many reasons, one of the more trivial being that it will not do what its proponents claim. In rushing to welcome investment from companies to exploit new ideas and discoveries, scientists seem to have forgotten or are unaware that most of our current hotshot ideas will in time prove to be wrong, irrelevant, or unimportant. Science is mainly in the business of invalidating the latest concepts. So why the rush to apply them?

But I have much deeper reasons for objecting to the industrialization of the university. The essence of an academic community is the free exchange of ideas, a sharing of knowledge. The formation of private companies within universities and with their faculty runs counter to this spirit. Private companies encourage a destructive kind of competitiveness that can be petty and mean. Secrecy becomes a priority when patenting ideas is a primary goal. And the lure of profit can result in both shoddy science and a narrow focus that ignores broader questions of social responsibility and the effect of new technologies.

My most serious concern is with the vital role of the academic as both critic and source of knowledge for society. Without an ax to grind, the scholar is in a unique position to provide a balanced point of view and has data to back him or her up. During the Vietnam War, two of the most visible activists among scientists were MIT's David Baltimore (who later earned a Nobel Prize) and Harvard's Mark Ptashne. They were critical of companies like Dow Chemical and Monsanto for their production of napalm, defoliants, and tear gas. Today, both Baltimore and Ptashne have their own biotechnology companies, and Dow and Monsanto are heavily involved in biotechnology. Do you think for a minute that Baltimore and Ptashne would be as critical of those industries today? Not on your life.

In the seventies, after the Arab oil embargo, I was involved in a film on the massive deposit of oil in Alberta's Tar Sands. At the time, with oil prices skyrocketing, there was talk of perhaps ten more oil extraction plants as big as or bigger than Syncrude. Each would produce at least fifty tonnes of sulfur dioxide a day. That's a lot of acid rain. So we tried to find a university ecologist in the area who would speak to us on camera about the environmental consequences of such development. We were unsuccessful because no one wanted to jeopardize his grant from the oil companies! Yet it is precisely for that knowledge that society supports such experts in a university.

I don't deny a role for university faculty in the application of new ideas. Our top-notch people are Canada's eyes and ears to the world's research, and good people will have ideas that can eventually be exploited. But the deliberate and urgent push to economic payoff distorts scholarship within the university and subverts its thrust to the will of those who have the money. Profit and destruction are the major reasons for the application of science today, whereas environmental and social costs are seldom seriously addressed. That's why we need scholars who are detached from those applications.

I remain a faculty member at UBC, and because I care so much for the university I am compelled to speak out in criticism. Tenure confers the obligation to do so.

I don't condone but can understand why university scientists, who have been underfunded for so long, are welcoming the Faustian bargain with private industry. But I fail to comprehend why philosophers, historians, and sociologists, who should know better, are acquiescing so easily.

The headlong rush to industrialize the university signals the implicit acceptance of many assumptions that have in the past been questioned by academics themselves. For example, free enterprise, like most economic systems, is based on the unquestioned necessity for steady growth—growth in GNP, consumption, and consumer goods.

Steady incremental growth within a given interval is called "exponential growth," and any scientist knows that nothing in the universe grows exponentially indefinitely. Yet economists, businesspeople, and politicians assume that the explosive increase in income, consumer goods, and GNP (and inflation) of the past decades must be maintained to sustain our quality of life. Historians know that this growth is an aberration, a blip that must inevitably stop and reverse itself. But how can the fallacy of maintainable exponential growth be seriously challenged when the university is busy selling the myth that it can help maintain such growth?

Scholars in universities represent tiny islands of thought in society. They are sufficiently detached from the priorities of various interest groups like business, government, and the military to point out flaws in our current social truths. But by focusing on issues that are socially relevant or economically profitable, we lose sight of the broader context within which that activity falls; we forget history; we become blind to environmental and social costs of our innovations.

In the U.S., a significant portion of the budgets of universities such as MIT, Harvard, Cal Tech, and Stanford now comes from private investment. This has split their faculties in debate about whether there should be such close ties with private enterprise. But whereas those institutions are private, Canada's major universities are all publicly supported. Yet there has been little debate in Canada about the imminent industrialization of academia. The activity and knowledge of our university scientists is paid for by the public and should be available for their benefit, not hidden behind a curtain of classified information, profit priorities, or patent secrecy. Academics who accept grants or investments from the military or the pharmaceutical, forest, and computer industries, for example, will be reluctant to jeopardize that support by criticizing those industries when necessary.

There is another consequence of the increased industrialization of our universities that originates in the mentality of scientists themselves. Among scientists there is a hierarchy of position that is directly correlated to grant size and continued research output. Scientists have to keep their "hand in" to maintain status and credibility with their peers. Anyone who decides to look at a wider range of social, environmental, or ethical matters, instead of focusing with tunnel vision on specific problems at the cutting edge of research, loses status in the scientific pecking order. Nobel laureates such as George Wald of Harvard and Cal Tech's Linus Pauling and Roger Sperry, who have become social activists and critics of some areas of science, are often referred to disparagingly as "senile," "over the hill," or "out of his area." As university scientists become bound to private enterprise more tightly, their horizons will be restricted even more and they will be far less patient with those who raise social and ethical questions about their work.

Let me be specific by considering one of the hottest areas of applied science—biotechnology—genetic engineering of organisms for commercial purposes. Biotech companies have been sprouting up on campuses like mushrooms. In a number of international meetings held at universities to discuss the future of biotechnology, none has seriously considered the potential misuse or hazards of the technology. Surely an academic community of scholars who maintain an arm's-length relationship with vested interests of society should be expected to raise those questions. Who else will do it?

One of the claims made to encourage greater investment in biotechnology is its potential to "feed the world's hungry." It is a self-serving, shallow justification. Starvation on this planet is a consequence more of political and technological factors than a shortage of food. Even if it weren't, the exponential growth of our species' numbers, which has already doubled the global population twice in the past century, will far outstrip any increase in food production brought about by biotechnology. Scientists anxious to justify their research for more support will resist such objections.

North Americans should be wary of the uncritical push to increase the links between university academics and private industry, because there are unacceptable "costs."

Live by the Box, Perish by the Box

THE FUTURIST BEING INTERVIEWED ON THE RADIO TALK SHOW ECSTATically extolled the benefits of the coming electronic revolution. After discussing the wonders of shopping, playing games, ordering movies, and checking the stock market, all from home through a computer and a television set, he gushed, "It sure beats having to put up with grumpy clerks, crowds, bad weather, and traffic jams." The tone of his remarks seemed to suggest that reality, like weather and other human beings, is a nuisance compared with the kind of controlled world we can now access through our TV sets.

The electronic revolution is being touted as offering a limitless variety of titillation and experience. The marriage of computer and telecommunication gives us "virtual reality," which, its proponents boast, is even better than the real thing.

Television is already the most pervasive and powerful medium of communication and information today, and it brings us more and more of our history lessons, values, priorities, and knowledge about the world. And the medium does provide astounding images few of us can ever experience in person—a view of war from the tip of a Patriot missile, close-up glimpses of Mars from a space vehicle, an intimate portrait of a patient's intestinal polyps, daily blow-by-blow skirmishes within a dysfunctional family.

Advances in the technological side of television have been breathtaking. They can be seen in their most impressive state on every broadcast of a sports event or in computer-animated commercials. From a viewer's perspective, television is better than reality; it's faster, more intimate, and clearer than real life. As we rush toward a 500-channel universe, we are told that television also has unlimited educational potential—but education about what?

Not long ago, the television set was referred to as the "boob tube," a pejorative expression that reflected the perceived lack of intellectual content of its transmissions. Not anymore. The television set will be the central component of the universe of virtual reality and the much-touted information highway. All but forgotten are those nettlesome questions about the real lessons being acquired from this electronic world by the viewing audience.

In real life, nature is exquisitely complex and diverse, but for television it has to be jazzed up because the pace of the natural world is too slow for the viewer conditioned to a constant stream of changing images. Consider what goes into a typical nature program. A wildlife photographer may spend months patiently waiting for a shot of a lifetime, one seldom seen by another human being. Incredible shots, such as those of a large mammal giving birth, avoiding a predator, finding food, or playing, are edited together to make a fast-paced program chock-full of great sequences. They can't help being powerfully moving and evocative. Yet the final impression is often more like "Animals Do the Darnedest Things" than a genuine insight into their daily routines. Don't visit the Amazon rain forest or an Arctic island if you expect to see the riot of color, shape, and movement portrayed in nature programs on television. Shows like that aren't a reflection of reality, they are creations.

By substituting television pictures for the real thing, we are profoundly distanced from it. And through living with the artifice of electronic images, we find it easier to think human technology and control are supreme, the sheer inventiveness of our species having allowed us to escape the constraints of our biology. It is such thinking that enables economists to reach the absurd conclusion that since 97 percent of the American economy is not directly dependent on climate, global warming will have little impact on the economy and the cost of any preventive measures will far outweigh the benefits.

Humanity's greatest need today is a restored sense of connection, interdependence, and love for other species. That can come about only by our experiencing directly with our bodies the vastness of the world around us and out there. We have to feel the heat and cold, smell the aromas and stink, savor the taste and the texture of real living things. We need to appreciate the ebb and flow of real time on geological, evolutionary, and biological scales, not the fragmented, disconnected, sped-up images that assault our senses through the box.

Most urban dwellers today feel uncomfortable, even frightened, by the unfamiliar surroundings of the real world that is our home. The futurist was dead wrong. The titillation of virtual reality and information superhighways is superficial and fleeting and renders us even more susceptible to the dangerous conceit that we no longer need nature.

A Humbling Message of Ants and Men

HARVARD'S EDWARD O. WILSON IS A WORLD-RENOWNED AUTHORITY on the variety of living organisms occupying the planet, an expertise based on a lifetime consumed with collecting and studying ants. Wilson's book *Naturalist* is an autobiography that informs us of how the wonders of the natural world became the formative focus of his childhood and persist as the source of his creativity and environmental activism. He emphasizes that nature has been the wellspring of our biological and social origins, provides the biophysical underpinnings of our lives, and inspires us with wonder and endless mysteries to ponder.

Wilson's story raises questions about our rush to lure students into science by stocking schools with expensive, glitzy interactive CD-ROMs and computers. The hi-tech machines carry an implicit worship of human technology and ingenuity, at whose altar much of the global environment has been sacrificed.

Wilson grew up in the Deep South during the Depression, a shy child whose parents divorced when he was seven. Yet his lost family life was compensated for by the education he received in the swamps of Alabama and on the coastal beaches of Florida. The very first memory he recounts is of a scyphozoan, a giant jellyfish, in the Gulf of Mexico. Like that of many scientists, Wilson's research career grew out of his childhood fascination with nature. "A child comes to the edge of deep water with a mind prepared for wonder. He is like a primitive adult of long ago, an acquisitive early *Homo*.... [H]e is given a compelling image that will serve in later life as a talisman, transmitting a powerful energy that directs the growth of experience and knowledge. He will add complicated details and context from his culture as he grows older. But the core image stays intact."

These days, acclaimed educational TV programs like *Sesame Street* perpetuate an accelerating information-and-sensory assault that precludes time for reflection or contemplation. Time is priceless. As Wilson writes:

> Adults forget the depths of languor into which the adolescent mind descends with ease. They are prone to undervalue the mental growth that occurs during daydreaming and aimless wandering. When I focused on the ponds and swamp lying before me, I abandoned all sense of time. Net in hand, khaki collecting satchel hung by a strap from my shoulder, I surveilled the edges of the ponds, poked shrubs and grass clumps, and occasionally waded out into shallow stretches of open water to stir the muddy bottom. Often I just sat for long periods scanning the pond edges and vegetation for the hint of a scaly coil, a telltale ripple on the water's surface, the sound of an out-of-sight splash.

Sadly, opportunities to duplicate that kind of childhood experience have become increasingly rare as wilderness vanishes and more and more children grow up surrounded by a human-created environment of concrete, tarmac, and television. But even in the most developed urban areas, there is an opportunity to experience wild creatures. All we have to do is focus on the realm of the small: "They are everywhere, dark and ruddy specks that zigzag across the ground and down holes, milligram-weight inhabitants of an alien civilization who hide their daily rounds from our eyes. For over 50 million years, ants have been overwhelmingly dominant insects everywhere on the land outside the polar and alpine ice fields. By my estimate, between 1 and 10 million billion individuals are alive at any moment, all of them together weighing, to the nearest order of magnitude, as much as the totality of human beings."

Then Wilson offers this humbling thought: "If we were to vanish today, the land environment would return to the fertile balance that existed before the human population explosion.... But if the ants were to disappear, tens of thousands of other plant and animal species would perish also, simplifying and weakening the land ecosystem almost everywhere."

After a life spent reveling in the abundance and variety of Earth's life-forms, Wilson has an urgent message:

> We are bound to the rest of life in our ecology, our physiology, and even our spirit.... When the century began, people still thought of the planet as infinite in its bounty. The highest mountains were still unclimbed, the ocean depths never visited, and vast wildernesses stretched across the equatorial continents.... In one lifetime, exploding human populations have reduced wildernesses to threatened nature reserves. Ecosystems and species are vanishing at the fastest rate in 65 million years. Troubled by what we have wrought, we have begun to turn in our role from local conqueror to global steward.... [O]ur perception of the natural world as something distinct from human existence has thus also changed fundamentally.

Wilson has observed catastrophic changes in tropical forests since his first field trips to the forests of Cuba and New Guinea in the 1950s. Escalating human numbers, consumption, and pollution have altered the planet, and to Wilson, the most harmful consequence is a steep decline in biodiversity. Life's rich multiplicity constantly cleanses air and water, replenishes the soil, and re-creates biological abundance. Once a species disappears, it can never be re-created. However, "if diversity is sustained in wild ecosystems, the biosphere can be recovered and used by future generations to any degree desired and with benefits literally beyond measure. To the extent it is diminished, humanity will be poorer for all generations to come."

Wilson points out that the most frightening aspect of the current extinction crisis is the enormity of our ignorance of what we are losing. Although only about 1.4 million species have been named by scientists, estimates of their number range from 10 to 100 million, with 10 to 30 million the most common guess. There are only 69,000 known species of fungi out of an estimated 1.6 million. Arthropods (which include insects), the most abundant group of species, have at least 8 or 10 million members in tropical rain forests alone. There are also probably millions of invertebrates that live on and beneath the ground and the bottom of deep ocean trenches.

But of all organisms, the ones we know least may be bacteria, of which a mere 4,000 species are recognized worldwide. Wilson cites a study carried out in Norway that found between 4,000 and 5,000 species among some 10 billion individuals in a gram of forest soil. Almost all of the species had never been identified before. When the biologists looked at soil taken from a nearby estuary, they discovered another 4,000 to 5,000 species; most of them were different from the forest sample and also new to science.

By comparing the estimated rate of species loss today with the changes observed in the fossil record, Wilson concludes: "The number of species on Earth is being reduced by a rate of 1,000 to 10,000 times higher than existed in prehuman times." The annual loss of about 1.8 percent of tropical rain forest, home to more than half of all species on Earth, extinguishes or endangers perhaps 0.5 percent of species. Wilson calculates that if there are 10 million species in these habitats, more than 50,000 species may vanish each year. This is a very conservative estimate that ignores the effects of pollution and of competing exotic species in different parts of the world.

Species relationships that evolved over millions of years are being wiped out in a geological blink of an eye. Wilson predicts, "Unchecked, 20 percent or more of the earth's species will disappear or be consigned to early extinction during the next thirty years. From prehistory to the present time, humanity has probably already eliminated 10 or even 20 percent of the species."

Extinction is a part of the evolutionary process. In the past, there have been episodes of mega-extinction, the most recent of which wiped out the dinosaurs. Wilson points out that there have been five major extinction spasms over the past 550 million years. On average, it took about 10 million years for evolution to restore the species abundance and diversity lost in each episode. So the catastrophic loss of a species taking place in a single lifetime will never be made up during our species' existence.

I have heard it suggested that since human beings are a part of nature, whatever we do must also be regarded as "natural." It therefore doesn't matter if we instigate a wave of extinction. Wilson replies: "The vast material wealth offered by biodiversity is at risk. Wild species are an untapped source of new pharmaceuticals, crops, fibers, pulp, petroleum substitutes, and agents for the restoration of soil and water. This argument is demonstrably true ... but it contains a dangerous practical flaw when relied upon exclusively. If species are to be judged by their potential material value, they can be priced, traded off against other sources of wealth, and—when the price is right—discarded. Yet who can judge the ultimate value of any particular species to humanity?"

I was a guest at a meditation session for cancer patients who had ridden a roller-coaster of hope and despair after chemotherapy, radiation, and surgery. Many spoke of "truly living" for the first time. Almost all made reference to the importance of "being in nature," whether walking in a woods, strolling a beach, or resting on a farm or at the cottage.

Watch children respond to a wasp or butterfly. Infants seem drawn to an insect's movement and color, often reaching out to touch it. They exhibit neither fear nor disgust, only fascination. Yet by kindergarten, this enchantment with nature somehow gives way to revulsion as many children recoil in fear or loathing at the sight of a beetle or fly.

Edward O. Wilson believes nature's attraction for cancer patients and infants is a natural inclination. He has coined the term *biophilia* (based on the Greek words for "life" and "love") to describe what he believes is a genetically programmed psychological need humans have for other beings. In his book *Biophilia: The Human Bond with Other Species,* he defines biophilia as "the innate tendency to focus on life and life-like processes." It leads to an "emotional affiliation of human beings to other living things.... Multiple strands of emotional response are woven into symbols composing a large part of culture."

Wilson suggests the origin of biophilia lies in our evolutionary history, which "began hundreds of thousands or millions of years ago with the origin of the genus *Homo.* For more than 99 percent of human history, people have lived in hunter-gatherer bands totally and intimately involved with other organisms.... They depended on an exact knowledge of crucial aspects of natural history.... The brain evolved in a biocentric world, not a machine-regulated world. It would be therefore quite extraordinary to find that all learning rules related to that world have been erased in a few thousand years."

Wilson's ideas echo those of the late microbiologist René Dubos: "We are shaped by the Earth. The characteristics of the environment in which we develop condition our biological and mental being and the quality of our life."

But today most people in industrialized countries live in urban environments in which the rich tapestry of other living things has been drastically reduced. Wilson believes that the "biophilic learning rules are not replaced by modern versions equally well adapted to artifacts. Instead, they persist from generation to generation, atrophied and fitfully manifested in the artificial new environments into which technology has catapulted humanity. For the indefinite future, more children and adults will continue, as they do now, to visit zoos than attend all major sports combined."

The notion of biophilia provides a conceptual framework through which human behavior can be examined and evolutionary mechanisms suggested. *The Biophilia Hypothesis,* edited by Wilson and Stephen R. Kellert, compiles articles examining biophilia and its implications. The papers add up to strong support for the theory. For example, the architecture professor Roger S. Ulrich reports that "a consistent finding in well over 100 studies of recreation experiences in wilderness and urban natural areas has been that stress mitigation is one of the most important verbally expressed perceived benefits."

Yale professor Kellert says:

> The biophilia hypothesis proclaims a human dependence on nature that extends far beyond the simple issues of material and physical sustenance to encompass as well the human craving for aesthetic, intellectual, cognitive and even spiritual meaning and satisfaction ... a scientific claim of a human need ... deep and intimate association with the natural environment.... The degradation of this human dependence on nature brings the increased likelihood of a deprived and diminished existence.... Much of the human search for a coherent and fulfilling existence is intimately dependent upon our relationship to nature.

To Kellert, our need for nature makes evolutionary sense: "Discovery and exploration of living diversity undoubtedly facilitated the acquisition of increased knowledge and understanding of the natural world, and such information almost certainly conferred distinctive advantages in the course of human evolution."

In the end, Wilson believes biophilia adds a spiritual dimension: "The more we know of other forms of life, the more we enjoy and respect ourselves.... Humanity is exalted not because we are so far above other living creatures but because knowing them well elevates the very concept of life."

Infoglut and Its Consequences

ACCESS TO MORE INFORMATION IS SUPPOSED TO HELP US PERCEIVE reality and deal with the big issues that confront us. And by any criterion we want to apply, more information is available to us than at any other time in history. Information overload, however, may create problems as severe as a lack of information.

Books, articles, and television programs herald the arrival of the Information Age. Telecommunications and computers form a global network that has become a key to economies of the world. We are the global village foreseen by the late communications wizard Marshall McLuhan and trumpeted by such futurist authors as Alvin Toffler.

When I began my career in television in 1962, it was my perception that science, when applied by industry, medicine, and the military, was by far the most powerful factor shaping our lives. Consequently, I believed the public needed more information about science to make better decisions about important issues in their lives. I was completely wrong.

By every standard we wish to apply, North Americans already have access to more information than ever before, but we still seem incapable of coping with the immense issues of nuclear war, pollution, species extinction, and so on. The problem is that we are inundated with information. Most of it is worthless to the average person, but we have no idea how to wade through the morass and separate the meaningful from the trivial.

Scientists refer to a situation such as this as having a "low ratio of meaningful signal to background noise." We have to ask critical questions about the nature of this information: What is it? Who provides it? What does it mean?

Today, we have specialists who tell us what we think. They are called pollsters. They sample public opinion on all kinds of issues for provincial or state and even municipal offices. But what do they tell us? During the lead-up to the Québec referendum on the province's future association with the rest of Canada, the Parti Québécois agonized for months over the wording of the referendum question because they surmised that phrasing would strongly influence the outcome. The questions posed by pollsters are not simply objective searches for information. The way the questions are asked influences the kind of responses that will be generated.

In any poll, the greater the number of response choices for any question, the more difficult the analysis. Hence, questions with yes-no answers are preferred. But the fewer the choices, the less meaningful the answers. Surely when Canadians are asked whether we favor free trade with the United States, membership in the North Atlantic Treaty Organization (NATO), participation in the Strategic Defense Initiative, or strong environmental legislation, simple yes-no answers reveal almost nothing of the complexity of either the issues or the thought behind the responses.

Even increasing the number of categories to "definitely yes," "yes," "perhaps yes," "perhaps no," and so on doesn't add very much. Usually the answers are reduced to mere percentages or pooled categories (all the yeses—"definitely" and "possibly"—are simply lumped together as "favoring"). We vest a great deal of value in numbers that cannot inform us about the variety of responses, the concerns, and the qualifications that shape our answers.

Nothing better illustrates the relative meaninglessness of information when it is simply packaged as a number than the distribution of scores on IQ tests. Forget the question of what IQ tests tell us about intelligence, because that is a matter of great debate. It is a fact that when a number of people take an IQ test, their scores will fall in a distribution that approximates a bell shape. People fall on either side of 100, and the farther away from that mean score, the fewer there are. Those who score above 120 represent less than 10 percent of the population.

In a city the size of Montreal, there may be 100,000 or more people who would score higher than 120. But so what? What does an IQ score tell us about the enormous human variability that resides in all the people who share that number? They will have just as many stupid, mean, avaricious, generous, or caring individuals as any other group along that curve.

By reducing people to mere ciphers, we give the appearance of scientific objectivity, but there is little real informational content. Most of these numbers tell us nothing abut the complexities that make issues important; they hide human values, fears, or hopes, social injustices or impediments to progress. Today, much of what the information purveyors do is collect massive amounts of data, condense them, and subjectively decide what is important.

I once gave a talk to executives in the electrical industry. One complained of the proliferation of books, magazines, and articles in his area. When I asked him how he dealt with them, he simply snorted and replied, "Hell, I don't have time to wade through the stuff. I tell my people to give everything to me on two pages or less. I prefer graphs with colors." It didn't surprise me, but it gets rather scary to think that decisions involving thousands of employees or millions of dollars may be made on the basis of a two-page summary of a one-hundred-page report.

Consider Canadian Cabinet members in federal and provincial governments or U.S. members of Congress or state governments. They work under enormous pressure and have no time to reflect on long-term implications or to read books and scholarly articles. They depend on their aides to summarize material, but what is lost in the reduction?

Daily we are assaulted with commercials, articles, and reports proclaiming the wonders of the Information Age. But we don't spend much time asking what it's all worth. We get our information today principally from television. We watch TV in large blocks of time—it is the main way we learn about ourselves and the world. And it's easy to see why: television is an almost effortless way of being informed, it's pleasant, and pictures make a powerful impression.

But if we are to be well informed, we must understand the medium, its limitations, and its modus operandi. We must realize that television does not reflect reality. Television reports are artificial, just as movies are. We don't have zoom lenses in our eyeballs; we don't live in a world where we can "cut" from one place to another or from one time to another. Yet that is what television does, and our brains fill in all the spaces. Try watching a report of a single event, such as a pole vaulter clearing the bar or a theatrical performance. It can be an illusion created by the juxtaposition of several cameras or repetitions edited together into an apparently single continuous piece.

News reports themselves are fabrications. From the decision as to what is newsworthy to the availability of a crew, the cost of doing the report and deadlines for editing and broadcast, stories sometimes are made or broken for reasons that have nothing to do with the significance of the events themselves. Once something is judged worthwhile (and that is often arbitrary and subjective), the item's length may be determined as much by whether an interview subject is articulate, or by the kinds of supporting visual material available, as by the inherent merits of the story.

And most important, if the researchers, writers, and reporters are upper-middle-class white males, chances are their unconscious social and cultural biases and values will color the way the report is finally presented. I say all this not to denigrate the medium but to point out that there is no such thing as objective reporting on television and people should understand that as they watch.

When I began my career in television in 1962, it was my conceit to think that through quality programming I could educate the public and raise its general awareness of the kinds of issues I was interested in. A quarter of a century later, *I* have been educated about the severe constraints in this endeavor. The problem is one of sheer *volume* of information: there's so much to absorb, and we no longer have a single coherent framework of values and perspectives by which to assess the information.

Change is what characterizes our lives—all about us, change is the one dependable feature of our social landscape. And because of rapid change and turnover, we acquire bits of information and quickly sift through them, retaining only those that happen to strike a chord. So often we repeat a "fact" that we justify with "I read somewhere that ... " or "I saw on TV that ... " without regard to the quality of the source. If the *National Enquirer* is viewed as being equally as credible as the *New York Times,* then information becomes totally devalued.

I am constantly surprised when in letters, phone calls, or personal encounters I am credited with all kinds of reports from programs I had nothing to do with. Often I am told, "I saw on your show ... " but it is a story we have never covered. People do not watch TV with concentration; there are interruptions—receiving phone calls, getting a snack, putting the children to bed. We tune in and out, watching when something interesting comes up and drifting off at other times. By the time we go to bed, the contents of a four-hour viewing block may be completely mixed up, and it is easy to assume that a snippet remembered from one show was actually seen on another.

We do it with other media as well, retaining little nuggets of information. This came home to me once when my wife, who teaches expository writing to science students, brought home a copy of the *National Enquirer* that she had used in a class exercise. I perused it with interest and found it was full of fantastic stories on medicine (woman inseminated by robot) and science (Soviets capture UFO crew). Three days later at work, I found myself mentioning that I'd read that there are toxic chemicals in the nipples of baby bottles. Midway through my recitation, I realized with a shock that I was repeating a story from the tabloid!

So while we have access to more information than ever in history, "infoglut," rather than educating and informing us, can make life much more complex and difficult. If we are going to take advantage of information, we have to understand the media and their limitations.

We need to assess the source of information. We have to believe in our own ability to judge the pros and cons of controversial issues and to demand access to primary information. That is precisely what scientists do in their profession. They are skeptical of any new claims, demand to see the evidence for themselves, and are confident enough to trust their own judgment about the worth of new claims. It's an attitude that the general public should find of great value in a time of informational overload.

Misusing Language

CURRENT PRACTICES IN FORESTRY ARE ENCOURAGED BY THE USE OF language. On a phone-in radio show about the controversy over whether to log the Stein Valley in British Columbia, Pat Armstrong, the paid spokesperson for the logging interests, defended "multiple use" of the watershed. (Incidentally, all but one participant on the hour-long program opposed logging the valley.) When one caller referred to the "devastation" and "destruction" caused by clear-cut logging, Armstrong bristled and said he refused to enter a discussion involving language like that.

Armstrong had correctly recognized the power of language. The words we use reflect and shape our attitudes and values. But the forestry industry that Armstrong defends so vigorously is itself a perpetrator of illusions created by the use of words. Foresters refer to "decadent" forests, meaning the stage in the evolution of a forest at which many economically valuable trees are approaching the end of their lives and will die. The implication is that they should be "harvested" to avoid a "waste"; yet for millions of years, forests have matured and changed while supporting a multitude of organisms.

Our Forest Services apply herbicides to forests to kill "weed" species of trees. Again, the word suggests a type of tree that is a pest or has no use. We hear of "thinning" or "culling" trees, and the cutblocks are referred to as "tree farms." Foresters talk about "plantations," "crops," "standard forestry practices," and "managing" wilderness areas. All these words have their roots in agriculture and carry the implication that forestry is simply the farming of trees.

We should not forget that agriculture is a sophisticated activity that is over ten thousand years old. Even though it has a long history, we have instituted such devastating practices as the heavy application of chemical pesticides and herbicides (thereby poisoning workers, consumers, water, and the soil), planting of vast acreages of single crops of uniform genetic makeup (which are extremely vulnerable to disease and create dependence on artificial fertilizers and other chemicals), and exposing much of the topsoil to erosion. In contrast, modern forestry is in its infancy and has never received the kind of financial support to maintain a top-notch, productive scientific community. The "crops" that foresters manipulate are not domesticated plants, and they must grow in uncultivated areas for decades. So the words used to imply that forestry has a similar basis as agriculture perpetuate a delusion.

Constant repetition of words and ideas reinforces a belief in their validity until they are assumed to be factual. A letter (*Globe and Mail,* June 29, 1988) from Dave Parker, a forester, who is now B.C. minister of forests, provides an illustration. He defends destructive clear-cut logging as:

> an accepted practice which follows the principle used by nature: clearing areas to regenerate forests. The difference is that nature uses insects, disease and wildfire. Nature's technique of removing hazardous material, controlling insects and disease and preparing a site for new forests by fire is emulated in forest management through prescribed burning....
>
> In British Columbia, we manage our forests on the basis of integrated resource management—considering all demands for use of the forest, whether those demands are for timber harvesting, recreation, preservation, and / or wildlife habitat, to name a few. In every forest management plan, all aspects of the ecological makeup of an area are considered.

A forest is a complex ecosystem made up of numerous organisms, many of which have not yet been identified and whose behavior and biological characteristics have not been defined. We have only the most superficial description of the inhabitants of a forest and almost no understanding of their interaction. How, then, can we consider "all aspects of the ecological makeup"? One thing that we have learned is that a major characteristic of wilderness areas is diversity—great variation in species composition, numbers of organisms, and the genetic makeup of each species. "Natural" forests do not grow back the way we reforest by planting uniform seedlings at regular intervals. Nor do we emulate nature when we apply pesticides, herbicides, and fertilizers to the soil, clear-cut huge tracts, and destroy the waste by burning. Communities of organisms coevolve over long periods and maintain a balance through natural selection. The immense machinery involved in clear-cut logging destructively churns up the soil. The operation drives wildlife from an area.

We often forget that soil is a living community of organisms, not just dirt. Each forest floor has its own distinctive accumulation of organic material from countless generations of plants and animals that have lived in an area. In British Columbia, much of the logged area is on steep slopes where the topsoil is thin and there is much rain. Much of the soil is quickly lost before a new group of trees can take hold and cling to the mountainsides. It is the height of self-delusion to believe that clear-cut logging imitates nature.

Of course, trees, plants, and animals can grow back where forests have been clear-cut. After all, about ten thousand years ago, all of Canada was buried under an immense sheet of ice and there were no forests. But it took all those ten millennia to create the natural treasures that we now enjoy, and it is a conceit to think that we know enough to act so wantonly and then re-create forests as if they are tomato or wheat fields. To suggest clear-cut areas are equivalent to parts of forests affected by natural fires, insects, and diseases reflects a failure to recognize the destructive nature of human practices. In the long-term interests of the forests that all of us want to share, we should stop misusing language to cover up our ignorance and inadequacies. This is not inconsequential: whole ecosystems are being destroyed under the impression that we can "manage" our resources.

The Really Real

TO CHANGE PUBLIC AWARENESS AND ATTITUDES, IT IS NECESSARY TO inform and educate people, and the best way to do that is through the media. But if individuals must take responsibility for judging the information that assaults them, they must understand the nature of information itself.

Television has great power; it delivers evocative images in vivid color and motion. These images are a major part of our experience of the world today. Our reality is based on experience that we sense through our nervous systems, and nothing is more convincing than what we see with our own eyes. It's hard to believe that those impressions are incomplete.

But reflect on this: our eyes can only detect electromagnetic radiation within a very limited range of visible wavelengths. Ultraviolet and infrared, which lie on either end of our detectable range, are invisible to us yet quite obvious to bees or rattlesnakes, respectively. Similarly, a dog experiences a very different world from ours through its nose and can sense a reality that to us is unnoticeable.

The same limitations apply to our senses of taste, smell, hearing, and touch. And not only do we detect just a fraction of all potential information available, but we alter a lot of the input we do get through the filters of our own preconceptions and experiences. Thus, a Stone Age person who sees a television program for the first time has the same visual and auditory impressions we do but nevertheless sees and hears something very different. It's not surprising, then, that it is often difficult to get two people present at the same event to recount it exactly the same way. Reality as we describe it is incomplete, a subjective construct of the human brain.

Scientists know reality in another way. It is based on a different mode of knowing and one that frequently flies in the face of our senses. To a scientist, an atom, a black hole, or the fifth dimension can be every bit as real and tangible as a rock. Laws of probability are real as well: we know that a healthy octogenarian who has smoked two packs of cigarettes a day since his teen years doesn't disprove the relationship between lung disease and smoking. Many find this difficult to grasp. There can be a blurring in the distinction between the statistical and the anecdotal because individual stories seem far more real and concrete.

I saw this blurring happen when I was on a committee on science education for the Science Council of Canada. One part of the study was a detailed profile of science teachers and science course offerings across Canada. A massive amount of information was accumulated, analyzed, and presented in graphs and tables. At the same time, case studies were carried out. Researchers visited randomly selected schools and observed the nature of teacher-student interactions in classrooms. What was striking was that the impact of the case studies far outweighed the statistical analyses. The case studies were rich in anecdotes and real human situations, whereas the analyses presented dry, impersonal numbers, which nevertheless gave a more accurate profile of the situation.

The same thing happened in a one-hour television broadcast several years ago on the CBC. It looked at the current state of cancer treatment and considered whether the "cancer establishment" had a vested interest in *not* solving the puzzle of cancer. Part of the program concerned the banning of the controversial drug laetrile in Canada. Scientist after scientist was interviewed and offered overwhelming evidence that laetrile was completely ineffective.

Then, at the end, the Hollywood actor Red Buttons described his wife's battle with terminal cancer. She had received chemotherapy and radiation therapy and doctors had concluded that there was no hope. Buttons went on to say that he took his wife to Germany, where she was treated with laetrile and was now considered cured.

Now any scientist knows that such a report does not *prove* a thing—

remissions may happen for all kinds of reasons, but one case does not allow us to pick out the factor responsible. Yet the power of the anecdote was irresistible—with that one story, all the scientific evidence seemed insignificant. Cancer experts were furious, and I don't blame them.

You can see the power of anecdotal reports on news shows when reporters discuss an issue and then turn to person-on-the-street comments. (I've done a lot of those "streeters," and there is nothing significant in them as a sample of opinion.) If the report includes two very strong pro comments and one weak anti position, the whole story is completely slanted. Yet those interviews are nonrandom samples that are further selected by the producers of the show to have an impact.

Although we have to distinguish between anecdotal material and quantifiable scientific data, we must never forget that science itself is carried out by human beings who have all the perceptual baggage of their society and their personal experiences.

Ever since the ancient Greeks posed the question "What is reality?" philosophers and scholars have pondered it through the ages. Far be it from me to tackle the question, but I can make some observations as a scientist and journalist.

Our enormous and complex brain gave us behavioral flexibility; we can think in abstractions, innovate, synthesize, project consequences of present actions into the future, learn, remember, and share knowledge. Other species may have some of these capabilities, but the output of our brains is unique on this planet. We are no longer bound by the restrictions of our neural circuitry; we have created machines that are incredibly faster, more reliable, and more efficient to bolster the output of our brains.

Through the 50,000 to 100,000 years of our species' existence, we have scarcely changed anatomically or genetically—Cro-Magnon man would not be noticed on the streets of New York, Toronto, or Sydney if he were dressed in modern garb. Yet in the millennia since *Homo sapiens* appeared, that brain has generated the enormous outpouring of language, song, poetry, art, and civilizations around the planet.

But if the sensory apparatus of all people is essentially the same, do we not begin with a common base of sensation of the world around us? After all, light waves focused on the retina elicit the same electrochemical impulses along neurons of all people. Aren't time, space, and sound experienced through the same nervous system? Here is where this remarkable organ begins to surprise us.

If you have ever experienced a sudden life-threatening situation or spent an agonizing wait for a special occasion, then you know that time expands or contracts depending on your "state of mind." Try going to a movie or watching a television program with someone and then discuss it later—it's amazing to compare notes and observe the differences in what is memorable or significant.

In my opinion, one of the most profound films ever made is the Kurosawa classic, *Rashomon*. In it, a woman and her husband are captured by a bandit who rapes the wife, then kills the man and escapes. When he is caught and brought to trial, the film flashes back to the events through the eyes of the woman, the bandit, the ghost of the dead husband, and a man who chanced to see it from the bushes. Each story is completely different!

Each of us, however closely we may share a culture and background, is a unique combination of genes and experiences that shape the perceptive capacity of our brains. We are constantly filtering out our experiences and creating a reality that is in essence an edited version of events.

In the film *The Big Chill*, there is a memorable scene in which one of the characters is accused of a rationalization.

"Of course," he answers, "rationalization is the most important thing in life."

"More important than sex?" he is asked. He replies, "Of course. Have you ever gone for three weeks without a rationalization?"

I agree. We are the great rationalizers—after a violent argument, both parties to it will reconstruct the events to cast themselves in the best light. My point is that all people feel that their view of the world, which has been shaped by their culture and experiences, is real. But put people from India, Botswana, and England in a similar set of circumstances, such as a serious illness, birth, marriage, or death, and you will witness reactions that are truly worlds apart.

Scientists regard science as a unique way of learning about the world because it provides a way to reproduce events in time and space and thus provides a picture of objective reality. But the history of science puts the lie to that. Scientists are first of all human beings, as subject to the limitations of cultural bias as anyone else.

So it should not be surprising that Charles Darwin, a member of the British aristocracy, educated in a period of slavery, colonialism, and mercantile expansion, should have described his great insight—evolution by natural selection—in terms of struggle and selection of the fittest. Today, cooperation and sharing are also seen as components of evolution. Geneticists at the turn of this century used their science to provide a biological basis for the "inferiority" of blacks, the poor, homosexuals, criminals, gypsies, and many others. Today, geneticists realize that science cannot make such evaluations. Neuroanatomists "proved" that white males had larger intellectual capacity than white females, who had a higher capacity than blacks and Native people, notions that are rejected now.

The list is long. This is not to say that scientists are bigoted or evil, only that they cannot escape their own as well as their society's preconceptions and social values. Nor is there any indication that scientists are any more enlightened or objective today than they were in times past.

Unfortunately, too few scientists or members of the general public are aware that data are not just data, that cultural framework, values, and assumptions affect the kinds of problems that are perceived, the way experiments are carried out, and the interpretations of the results.

Even when we understand the pitfalls of scientific "objectivity," we are still stuck with the perceptual limitations imposed by our senses. But our senses may no longer be enough to survive the hazards of the radically changed world of the last few decades.

Two grapes tainted with cyanide were enough to bring a multimillion-dollar Chilean agribusiness to its knees. The discovery of those two fruits must represent an incredible achievement by the customs office. On the basis of an anonymous phone tip and those two poisoned fruits, several governments, including those of Canada, the United States, and Japan, immediately imposed a ban on all Chilean fruit and vegetables. Yet those same governments maintain a remarkable tolerance of wide public exposure to the chronic presence of PCBS and potential hazards such as Alar in North American–grown food. We are able to respond heroically to an immediate perceived crisis far more readily than we can to much more significant global problems. Why?

That is the question posed by two Stanford University professors, psychologist Robert Ornstein and ecologist Paul Ehrlich, in a provocative book, *New World, New Mind*. They remind us of numerous contradictions in human behavior. We can be deeply moved by reports of a child trapped in a well in Texas while ignoring the deaths of twenty million infants a year from diseases that are preventable. We are mesmerized by images of three whales trapped in ice off Alaska but seem unconcerned about global extinction, which is now estimated to claim more than *two species an hour.* Threats by a small band of terrorists create widespread fear, yet we tolerate the camage on North American highways that claims over fifty thousand lives a year and the continued enlargement of an already massive global arsenal of nuclear weapons.

For Ornstein and Ehrlich, the explanation of these inconsistencies lies in our long evolutionary history and the speed with which changes now occur. We detect events happening in our surroundings through the apparatus of our senses. Our reality is delineated by the range of those senses. But in spite of such limitations in our sensory input, we are able to respond magnificently to our environment. Ornstein and Ehrlich point out that if we are in a cave and the shadow of a bear crosses the entrance, or if we are standing on a branch and we hear it crack, we are able to respond immediately and take protective action.

As well as coordinating the input from our senses and sending out appropriate responses, the human brain also invented an idea called *future.* Because we recognize that what we do today can affect what happens in the future, our species is uniquely able to select from a number of options to maximize survival. We have deliberately chosen a path (albeit strewn with dead ends and wrong turns) into the future, and it has worked so well that we are now the most ubiquitous and numerous large mammal on the planet.

But our enormous increase in numbers, coupled with an acceleration in technological muscle power conferred by science, has created a "new world" in which our "old minds" are no longer adequate. For most of history, small groups of hunter-gatherers were not even aware of the existence of others of their kind in distant places. Their world was circumscribed by how far they could communicate and travel and where they could find food. When human numbers were small and technology simple, their environment could easily absorb and recover from the impact of human activity. Human beings only reached a billion early in the past century, yet we have already passed five billion and appear on the way to doubling in fewer than another fifty years. Although our numbers are now vast, we continue to respond to immediate and personal crises rather than to those that affect our species as a whole. Whereas once the hazards to be avoided were immediate and tangible, today they are more abstract and lie years away.

Many dangers we face are now beyond the ability of our sense organs to detect. We can't tell that there are pesticides in our food, dioxin or PCBS in water, or an increase in low-level background radiation or radioisotopes. We are unaware of the ozone layer or changes in it or an increase in CFCS or carbon dioxide in the atmosphere. Acid rain doesn't register with us physically. And since we have adapted to respond to threats that we can "feel," we find it difficult to take the new perils of today's world seriously.

Ornstein and Ehrlich believe that the constraints of human perceptions that affect our priorities and actions are not immutable consequences of our genes. They believe that it is possible to develop a "new mind" through a better understanding of the global community of organisms of which we are a part and of the finite resources that must be shared and recycled by all forms of life. The massive response of governments to two poisoned grapes should be an occasion to reflect on what we must do to counter the *real* hazards of the new world.

Television's Real Message

INCREASINGLY, THE ELECTRONIC MEDIA, ESPECIALLY TELEVISION, play a major role in providing us with experience and knowledge of the world around us. Although reporters and producers aim for balance and objectivity, they cannot escape the cultural filters that determine the way we see the world. At the very least, we have to recognize these built-in biases.

Television's power comes from its heart-clutching shots of devastating wars such as those in Iraq or Sarejevo. We watch cyclone victims in the deltas of Bangladesh in numbers that are beyond comprehension and watch people die before our eyes in Somalia.

But the images are ephemeral, flitting through our consciousness to be replaced by reports on the state of the economy or sports scores. So by the time we have turned off the television set to drive over to the local mall to go shopping, the suffering and misery of the unfortunate people in those distant lands have receded from our consciousness. Important events become trivialized because the news of the world is reported in the electronic media as a series of disconnected fragments.

Yet the most profound lesson of the global ecocrisis is that our lives are intimately connected to those events taking place in the Middle East, Africa, and the Indian subcontinent. The Arab countries are an international hot spot because the world's industrialized nations have not learned to live within their means and have become predators of the resources located in the Middle East. Stanford University ecologist Paul Ehrlich points out that this situation could have been avoided if the United States had adopted a population strategy to stabilize its population at the 150 million level when he was born. Then the United States would have remained completely self-sufficient in oil. In a very real way, then, the Kurdish refugees of the Gulf War can trace their plight to our way of life.

Although the media present "global warming" as if it remains scientifically controversial, scientists around the world are nearly unanimous in thinking that the threat is real and demands immediate action by governments of industrial nations. Each of us as individuals in Canada, the United States, Australia, and other developed countries contributes disproportionately to the increasing concentration of greenhouse gases through our lavish lifestyle. We squander gas in our cars—every liter (quart) of gasoline, which weighs 0.8 kilograms (1.76 pounds), burns and releases 2 kilograms (4.4 pounds) of carbon dioxide into the atmosphere. A large, healthy tree takes a year to remove 9 kilograms (20 pounds). And we are using a lot of gas while cutting down large numbers of trees.

The rise in sea level as ocean water expands will pose a tremendous hazard to chronically flooded lowland deltas in Bangladesh and Egypt and to the globe's thousands of coral islands. Tides and storms will increase in severity and unpredictability as weather patterns become erratic and create large numbers of environmental refugees. A sea level rise of only a few centimeters will be disastrous for all those who live on marine coastlines.

Drought and famine also result from global change in weather patterns. At the Geneva Conference on Global Climate Change in November 1990, scientists from Kenya and Nigeria told me that farmers in their countries depend on seasonal cues from nature (monsoon rains, blossoming of certain indicator plants, abrupt temperature changes) to tell them when to plant crops or begin the harvest. Now the farmers tell them that weather has become so erratic that those signals are no longer reliable.

The average Canadian consumes more than sixteen times as much as the average person in the Third World. Already, many of our health problems can be traced to pollution, lack of exercise, and overeating, which are a direct result of the way we live. Yet our relentless drive for even greater economic growth and consumption merely exacerbates environmental problems at home and around the planet. The vast quantities of resources required by our industries deprive poor countries, squeezing them between greater demands from growing populations and a crippling economic debt load.

Images on TV—like Bob Geldof's Live Aid—can arouse immediate action to help our fellow human beings. But we need to be aware of the global ecological context that can make sense of the disjointed fragments we see.

Today's television audience is overwhelmed by a technology undergoing explosive change—cable and dish antennae bring dozens of channels into homes where a viewer armed with an infrared zapper can graze through programs around the clock. Exactly what is being "learned"?

At the Wildlife Filmmakers' Symposium in Bath, England, in 1990, Bronx Zoo filmmaker Thomas Veltre suggested that "ethics are embedded in all technologies." He cited the chain saw as an example. It's not just a labor-saving tool; it "can alter a culture's entire relationship with the forest. Almost overnight teak and mahogany trees cease to be habitat. Instead, they become resources—cash crops waiting to be exploited. The need to buy gasoline and spare parts, even the saw itself, plants one firmly in an international cash economy."

Veltre believes that television delivers a message that is fundamentally opposed to an environmental conservation ethic, which is to "be restrained and cautious in our consumption of natural resources, see the world as coherent and interconnected, and be farsighted—looking decades or even centuries ahead." Television, in contrast, "encourages a culture to be impatient, incoherent, and shortsighted." Television is not just a collection of different programs. There is an effect of the sum total on the way we think and act. Thus, Veltre says, whereas "conservation encourages people to delay their gratification, to not consume today the resources they will need tomorrow ... television is impatient; it will not tolerate delay of gratification. The symbolic form of television is instantly accessible," unlike books, which require effort and commitment by a reader.

He goes on to point out that "conservation wants people to see the world as coherent and interconnected, that events happen in a context, and that actions in one place have consequences in others.... Television's approach to the world is incoherent. Events in one time and place have no connection to any other time or place.... Television is free to supply all the images it needs, as fast as it can, in any order, from anywhere in the world....

The ultimate message of television is that nothing need be connected to anything, so long as the pictures are good."

Finally, Veltre indicates "conservation encourages a farsighted view, an informed historical perspective to help make decisions and ... when considering the consequences of those decisions ... an attention span for changes that happen slowly." In contrast, "television is by far our most shortsighted medium.... It builds an expectation of quick change ... and encourages a culture to ignore the long-term," thereby trapping that culture "by the tyranny of an ever-changing present." (I was struck by the accuracy of that statement when I heard a DJ on a rock station announce, "And now for a golden oldie from last year!")

Most people on the planet today have lived their entire lives since the end of World War II, an unprecedented and aberrant period of growth and change that the electronic media indoctrinate us into accepting as normal. As Veltre points out, "Television, which encourages one to think no further than the next commercial, can only attend to change that occurs instantly, or at best within the course of a single day. Things that change slowly, like the environment, cannot be televised.... A culture based on the idea of change at breakneck speed, with no regard for its past and no concern for its future, may be lost to conservation forever."

What are the solutions? Veltre recommends that Europe "resist the deregulation movement ... that four channels are more than enough." We should also "support any and all forms of programming which struggle against the ethic of the technology." Perhaps conservationists ought to "reconsider why they use television at all." Instead, he proposes the use of other traditional media, like books, "events and rituals, music, art, sculpture, perhaps even buildings." Thus, like the architecture of the Bronx Zoo, we should attempt to allow people "to encounter the beauty and wonder of life, and to return again in reverence to contemplate its mysteries."

To illustrate Veltre's claim that information is fragmented and disconnected physically and temporally from the rest of our lives, consider a single news report on a very ordinary day.

March 4, 1992, was a typical spring day in Vancouver—temperature over 10 degrees Celsius (50 degrees Fahrenheit), rain, and streets ablaze with blossoms. Listening to early-morning CBC radio's usual fare of short, fragmentary reports presented in a "balanced" and dispassionate way as "world news," I was struck by how little information they impart. The stories fail to provide a context within which to assess them and recognize their broader implications, such as the fact that the planet's ecosphere is being torn apart. Consider what we heard on that one typical morning.

Item 1: The multibillion-dollar megaproject Hibernia was supposed to be an economic boon to a depressed Newfoundland but received another blow. Petrocan president Wilbert Hopper announced that unless another investor is found to replace Gulf Canada Resources Limited, they, too, could pull out in sixty days. But what about the important questions? Do megaprojects really create long-term stability in local regions? Do local people get most of the jobs? Should we invest in expensive oil development that adds to global warming? What are the risks to a sustainable activity like fishing? There isn't time for such questions when news clips are only fifteen to forty seconds long.

Item 2: The Saskatchewan government is debating whether to open up more uranium mines. Proponents of the mines point to the economic potential during a severe recession, and opponents warn that uranium could end up in weapons. Unreported was the terrible radioactive contamination of water and fish in aboriginal lands. The uranium story could have made an important point: we must begin costing our activity in an ecologically effective way. Uranium isn't just about jobs and revenue but also about health, environmental pollution, waste management, and international nuclear weapons. We should be estimating cost from "cradle to grave," but that doesn't make the news.

Item 3: Prime Minister Brian Mulroney and Newfoundland premier Clyde Wells called for greater vigilance of the foreign fishing fleet that is plundering plummeting northern cod stocks. But how credible is it for politicians preoccupied with the economy and jurisdictional matters to act as if they understand ecology and have a well-thought-out plan of action? For twenty years there have been warnings that the northern cod are being overfished.

Item 4: An American pulp mill announced that it would not give in to European demands for nonchlorine-bleached pulp. Canadian companies apparently would like to do the same but know that government legislation will force compliance with those demands. It's never asked why staunch defenders of free enterprise and global competitiveness untrammeled by legislation cry foul when they are pressured through the marketplace. Instead, they accuse environmentalists of being irrational zealots, interfering with global competitiveness or participating in an unfair boycott. But when asked to justify emissions exceeding sustainable levels, they respond they are only doing what governments allow.

Item 5: Thirty-seven million dollars may finally be paid to citizens of Haida Gwaii (Queen Charlotte Islands) as compensation for the decision made in 1987 to preserve Gwaii Hanaas (South Moresby) as a national park reserve. We didn't learn why it is taking so long to establish the park, what the eventual role of the Haida and the non-Native people will be, or why the company that merely held logging rights to the Crown land and did not invest any money in those public forests has already been compensated with $40 million.

Item 6: Gypsy moths have been found in the Lower Mainland of British Columbia, and a government plan to spray BT, a biological pesticide, is being opposed by citizens. Such stories invariably present gypsy moths as a severe threat to trees. We don't hear whether spraying can ever eliminate an exotic arrival once it has a toehold. Nor are we told that insects are by far the most numerous group of animal species on Earth and that most of them are vital food for many animals and act as pest controls, pollinators, decomposers, and so on. Perhaps one in a thousand insect species is a nuisance to human beings, yet we undertake massive programs that will affect a wide spectrum of insects just to get at the tiny fraction that we don't want. Is that sensible management?

March 4 was another typical day for radio news and was, in fact, full of stories with profound ecological implications. If information were packaged and delivered by the media in a way that sets out the broader context and links stories into a cohesive picture, we would quickly recognize that the planet is in trouble and there is little indication of leadership or vision to deal with it. Only then can we begin to work toward solutions.

Virtual Reality

DOES TELEVISION DO THE JOB OF REPORTING ANY BETTER THAN radio? Surely documentary programming presents the reality of the world that is covered. After all, pictures never lie. That cliché is no longer true, however, as the line between photographs and computer-generated images is getting harder to discern. Imagine this scene from a nature documentary:

> Deep in the heart of the Amazon rain forest, a troupe of howler monkeys swings through the treetops, 100 feet from the ground, hooting like a football crowd as it passes. A sloth hanging from a lower branch is shaken from its lethargy and starts to climb slowly down the great buttressed trunk. Out of the shadowy recesses of the forest, two giant butterflies appear, huge, iridescent blue wings flapping in synchrony as they circle a brilliant shaft of sunlight.

The scene is meticulously accurate: all these animals live in that region of the Amazon; they are in the place they belong, doing what they normally do. But the documentary form has its own way of re-creating reality. This two-minute scene is selected from hours of footage that took days or even weeks to shoot. The magic of editing creates a flurry of activity in the forest, puts together a crowd of creatures in a place that in reality is mostly silent and still.

As the flow of information continues to increase, newspapers compete for our attention by compressing reports into single paragraphs. Radio and television stories chop "interviews" up into sound bites, leaving a personality or politician just time to say "I don't agree," or "There is no cause for alarm." News reports may range from fifteen to forty seconds, and an "in-depth report" might last for two whole minutes.

The overabundance of information is reflected in the tidal wave of brief fragments coming at us. According to Theodore Roszak, writing in *The Cult of Information,* a single weekday edition of the *New York Times* contains more information than the average person in seventeenth-century England would have encountered in an entire lifetime. In addition, computer processing speed has doubled every two years for the past thirty! As a consequence, information has become compressed. So between 1965 and 1995, the average length of a TV commercial shrank from 53.1 to 25.4 seconds, and the average news sound bite was reduced from 42.3 to 8.3 seconds! At the same time, according to *TV Dimensions '95* and *Magazine Dimensions '95*, the number of ads squeezed into a minute went from 1.1 to 2.4.

I was confronted with a striking confirmation of the compression of information in 1992 when we were preparing a program on the Earth Summit being held in Rio. By then, *The Nature of Things* had been an hour-long program for thirteen years. In the archives, I looked up the report done in 1972 on the UN conference on the environment held in Stockholm. The program was still a half-hour long back then, and to my amazement it contained three- and four-minute interviews with Paul Ehrlich and Margaret Mead. Today, we never put on an interview longer than thirty to forty seconds. And it was clear my expectations and interest span had changed, because I found the 1972 interviews slow.

Thus, the shorter, punchier accounts we now get are stripped of any kind of historical or contextual material that might make the news event more understandable in a larger sense. Each night a stream of unrelated stories gives us a glimpse of a fractured, puzzling, often frightening universe. There are battles being fought by mysterious factions in countries whose past is a blank. Violence and misfortune are the normal state of affairs, forest fires rage, trains crash, and deficits rise. Studies are reported on subjects we have never thought about, and politicians announce decisions without rhyme or reason. Modern life exists in a confusing, chaotic world that is much as it must have seemed before people were formally educated.

The Hidden Messages

THE MEDIA ARE FASCINATED BY THE NEWEST MEMBER OF THEIR group, the information superhighway, marveling at its apparently limitless potential. Cyberspace, virtual reality, the World Wide Web, five hundred channels of interactive television—what a brave new world this is to be. U.S. vice president Al Gore, a committed environmentalist, is also a strong supporter of the limitless benefits of the information superhighway.

But beneath the hype and techno-adulation, there are troubling questions. Theodore Roszak's book *The Cult of Information* raised some of them:

> For the information theorist, it does not matter whether we are transmitting a judgment, a shallow cliché, a deep teaching, a sublime truth, or a nasty obscenity. All are information.... Depth, originality, excellence, which have always been factors in the evaluation of knowledge, have somewhere been lost in the fast, futurological shuffle.... [T]his is a liability that dogs every effort to inflate the cultural value of information.... We begin to pay more attention to "economic indicators" than to assumptions about work, wealth and well-being which underlie economic policy.... The hard focus on information that the computer encourages must in time have the effect of crowding out new ideas, which are the intellectual source that generates facts.

Here is another take on the so-called benefits of information and the computer revolution from Allen D. Kanner and Mary E. Gomes, authors of *The All-Consuming Self:*

> Priority is being given to the technology necessary for around the clock interactive shopping. Television sets are being transformed into electronic mail catalogues. The goal is to allow viewers to buy anything in the world, any time of day and night, without ever leaving their living rooms.

Clifford Stöll, author of *Silicon Snake Oil: Second Thoughts on the Information Highway,* has been deeply immersed in the information network since its inception. His five modem-equipped computers are plugged into the info world; he surfs the Net regularly, and he loves his cyber-community. Nevertheless, he has profound misgivings about the technology:

> They [computers] isolate us from one another and cheapen the meaning of actual experience.... [M]ore than half of our children learn about nature from television, a third from school and less than 10 percent by going outdoors.... [N]o computer can teach what a walk through a pine forest feels like. Sensation has no substitute.

Curious students who are excited about learning will take to computers as readily as they do to literature, history, and science. But Stöll reminds us that

> isolated facts don't make an education. Meaning doesn't come from data alone. Creative problem solving depends on context, interrelationships, and experience. The surrounding matrix may be more important than the individual lumps of information. And only human beings can teach the connections between things.

We are surrounded by information and information technology. But they do not tell us what we need to know: how to live in balance with the natural systems of the planet. Ecological issues require a different kind of information: material we can use for long-term thinking, for seeing connections and relationships, and for acting in cautious, conservative ways.

Writing in the *Dominican Quarterly,* Philip Novak cautions that excessive information can have another very damaging effect: "Superabundant information is grand, until we understand that it can rob us of the peace that is our spiritual birthright. We have only recently realized our need to develop an ecological relationship with the natural world. Perhaps we must also realize our need for inner ecology, an ecology of the mind."

Computers have been pushed as one of the great revolutionary hopes for education, and Steven Jobs, the fabled cocreator of the Apple computer, has been one of the strongest advocates of this notion. But even he has had second thoughts, according to an interview he gave to *Wired* magazine's Gary Wolf in 1996:

> I used to think that technology could help education.... I've had to come to the inevitable conclusion that the problem is not one that technology can hope to solve.... Historical precedent shows that we can turn out amazing human beings without technology. Precedent also shows that we can turn out very uninteresting human beings with technology.

The reason that computer technology fails to deliver on its expectations is that whereas information can be accessed in vast quantities and with great speed, education is about sifting through it, making sense of ideas and what we perceive. No computer can do that, as Alan Kay, one of the pioneers in personal computing, testified before Congress in 1995:

> Perhaps the saddest occasion for me is to be taken to a computerized classroom and be shown children joyfully using computers. They are happy, the teachers and administrators are happy, and their parents are happy. Yet in most such classrooms, on closer examination I can see that the children are doing nothing interesting or growth-inducing at all! This is technology as a kind of junk food—people love it but there is no nutrition to speak of. At its

> worst, it is a kind of "cargo cult" in which it is thought that the mere presence of computers will somehow bring learning back to the classroom.

Or as Neil Postman said in the *Utne Reader,* "We have transformed information into a form of garbage."

Even a big booster of the information revolution like former U.S. president Bill Clinton warns us, "In the information age, there can be too much exposure and too much information and too much sort of quasi-information.... There's a danger that too much stuff cramming in on people's minds is just as bad for them as too little, in terms of the ability to understand, to comprehend."

In our infatuation with the seemingly wondrous possibilities the computer creates, we welcome it into the school only to find that it has become another vehicle for marketing opportunities. Marketing companies like Lifetime Systems design commercial packages to look like educational material, wrote David Shenk in a 1994 article for *Spy* magazine. The company points out the opportunities to prospective clients: "Kids spend 40 percent of each day in the classroom where traditional advertising can't reach them. Now you can enter the classroom through custom-made learning materials created with your specific marketing objectives in mind. Communicate with young spenders directly, and through them, their teachers and family as well."

The highly fragmented offerings in an ever increasing menu of television channels have been paralleled by a growing number of both news channels and news programs. Along with CNN, MSNBC, SKY, BBC World, and CBC Newsworld, we have the analogue of print tabloids in *Hard Copy* and *Cops.* As Shenk says:

> The news-flash industry supplies us with entertainment, not journalism, and as such is part of the problem of information glut.... Our fundamental understanding of Bosnia or the stock market is not going to change, no matter how many news-bites we hear about them. To actually learn about the subject requires not a series of updates, but a careful and thoughtful review of the situation.

As the quality of air, water, and landscape degenerate from the assault of human activity, more and more of our fellow human beings find themselves in sprawling megalopolises. In cities we are distanced from the natural world, spending more and more time in search of stimulation in shopping malls, electronic games, and television. In lieu of the experiences of the real world, we now have all of the gut-wrenching, adrenalin-rush, sensory overload of "virtual reality." The truly horrifying aspect of the cyber-world is that it appears to be *better* than the real world. After all, one can access the virtual world of sex and experience every kinky possibility without fear of contracting AIDS, feeling guilty, or being caught by one's partner. We can take part in a gunfight, lose, and live to shoot again; race a car, crash, and walk away; or get blown up or beaten to a pulp without pain or injury. Who needs a real dangerous world when all of our sensations can be zapped to the max without risk?

During the '70s, on a noon-hour television talk show, the host asked me, "What do you think the world will be like in a hundred years?" I replied, "Well, if there are still people around then, I think they'll curse us for nuclear power and television." The host did a double take and, ignoring my caveat about still being around, asked, "Why television?" My answer was as follows: "Well, Bob, you just asked me a pretty tough question. If I had answered, 'Gee, I'll have to think about that for a minute,' and then proceeded to think without saying a word, you'd cut to a commercial in less than five seconds. Because television cannot tolerate dead space. It demands instant response. There's no room for reflection or profundity. It's not a serious medium."

Reflecting now on that answer, I wouldn't change my assessment. However, that response was given years before the multichanneled universe of cable, which demands even faster, snappier programming to keep our attention. The one element that the real world of nature requires for us to experience it is *time.* So the more our children experience nature through television and films, the more they will be disappointed when they encounter the real thing. Wild things don't perform on cue. They are shy. Often they are active only at night or underground. Nature allows us only rare moments when there might be a flurry of activity. And the waiting makes the experience all the more satisfying when nature does reward us.

Are These Two Reporters on the Same Planet?

THE MEDIA THRIVE ON NOVELTY. WHEREAS DAILY NEWSPAPERS struggle for readers, tabloids flourish with lurid stories from outer and inner space. As competition for viewers intensifies with the proliferation of television channels, TV stories become shorter, kinkier, more violent, and more sensational than ever. All the while, our threshold for shock and violence rises.

Demand for increased titillation has changed the nature of documentary reporting. When I began television reporting in 1962, three- to four-minute interviews with articulate, thoughtful scientists were not at all unusual. Today they might get half a minute. Stories are shorter, punchier, faster, and slicker, but they are also shallower and provide less detailed historical and social context. In-depth reporting gives way to the exploitative and anecdotal. It's not surprising, then, that Paul Bernardo, Lorena Bobbit, and Tonya Harding received coverage far in excess of their global significance.

During the 1970s and 1980s, as public awareness grew over the ramifications of our lifestyles and technology on the planet, we were constantly shocked and surprised by the unexpected interconnections and consequences. Who would have believed that DDT sprayed on fields to kill insects would end up causing thinner eggshells for birds, or that the heavy industries of Pennsylvania would affect trees in Québec? The stories were alarming, and the media reports reflected it.

But it was inevitable that sooner or later, stories that once sparked shock and outrage would induce yawns. Nothing is more stale than yesterday's newspaper. Peter Desbarais, former dean of journalism at the University of Western Ontario, told *Maclean's,* "I tend to resist articles announcing some new environmental threat. I feel that I've heard it all before." He's right, although not in the way he meant it. Virtually all environmental problems can be traced to the same causative factors of rapidly growing human population, overconsumption, and excessive technological power. But whereas murders, wars, business failures, political crises, or sports finals are eventually resolved, the solutions to environmental problems are seldom simple and easy; they are complicated and require long-term attention. That doesn't make for good press.

The stories that do emerge often question the credibility of the environmental issues themselves. A spate of books, articles, and television programs have disputed the reality of the claimed hazards of global warming, overpopulation, deforestation, ozone depletion, and so on. Other stories are built around ever more frightening possibilities.

An article in the *Atlantic Monthly* by Robert D. Kaplan has galvanized both fear and denial. Entitled "The Coming Anarchy," the report paints a horrifying picture of the future for humanity. Kaplan suggests that the terrible consequences of the conjunction between exploding human population and surrounding environmental degradation are already visible in Africa and Southeast Asia. As society is destabilized by an epidemic of AIDS, government control evaporates, national borders crumble beneath the pressure of environmental refugees, and local populations revert to tribalism to settle old scores or defend themselves against fleeing masses and marauding bands of stateless nomads.

Kaplan believes that as ecosystems collapse, this scenario could sweep the planet, first in the Eastern bloc countries and then the industrialized nations. It is a frightening scenario built on a serious attempt to project the aftermath of ecological destruction. And it has generated a great deal of discussion and controversy.

Marcus Gee pronounces Kaplan's vision "dead wrong" in a major article in the *Globe and Mail* headlined "Apocalypse Deferred." Assailing "doomsayers" from Thomas Malthus to Paul Ehrlich and the Club of Rome, Gee counters with the statistics favored by believers in the limitless benefits and potential of economic growth. Citing the spectacular improvement in human health, levels of education and literacy, availability of food, and length of life even in the developing world, Gee pronounces the fivefold increase in the world economy since 1950 to be the cause of this good news. He does concede that "immense problems remain, from ethnic nationalism to tropical deforestation to malnutrition to cropland loss," but concludes that Kaplan has exaggerated many of the crises and thus missed the "broad pattern of progress."

Are these two reporters on the same planet? How could they come to such different conclusions? And what is the reader to conclude?

Kaplan believes what he saw in Africa and Southeast Asia was the beginning of a global pattern of disintegration of social, political, and economic infrastructures under the impact of ecological degradation, population pressure, and disease.

In contrast, Gee focuses on statistics of the decline in child mortality and the rise in longevity, food production, and adult literacy in the developing countries to reach a very different conclusion—things have never been better! Economic indicators, such as a rise in gross world product and total exports, indicate, he says, "remarkable sustained and dramatic progress ... life for the majority of the world's citizens is getting steadily better in almost every category."

Kaplan's frightening picture is built on a recognition that the planet is finite and that degradation of ecosystems by the demands of population and consumption has vast social, political, and economic ramifications. Gee's conclusions rest heavily on economic indicators. He points out the annual 3.9 percent rise in the global economy and the more than doubling of the gross output per person over the past thirty years. World trade has grown by 6 percent annually between 1960 and 1990, as tariffs have declined from 40 percent of a product's price in 1947 to 5 percent today. Yet all this time, the gulf between rich and poor countries has increased.

Gee skips lightly over such facts as Third World debt and the death of 22,000 children a day of easily preventable diseases. He admits the real threats of loss of topsoil, pollution of the air, loss of forests, and contamination of water, but he concludes there is little evidence that they are serious enough to halt or even reverse human progress. He even suggests the preposterous notion that global warming and ozone depletion "may cancel each other out."

Gee's outlook rests on a tiny minority of scientists who have faith in the boundless potential of science and technology to transcend the physical constraints of air, water, and soil so that a much larger population can be sustained. His final proof? The concomitant rise in living standards and population. But the relationship between changes in living standard and population growth is a correlation, not proof of a causal connection.

Gee quotes the "American scholar" Mark Perlman: "The growth in numbers over the millennia from a few thousands or millions of humans living at low subsistence, to billions living well above subsistence, is a most positive assurance that the problem of sustenance has eased rather than grown more difficult over the years." Even the World Bank, which is not known for its sensitivity to ecosystems or local cultures, is quoted as stating "the food crisis of the early seventies will be the last in history."

Gee relies heavily on Julian Simon, once an economic adviser to Ronald Reagan. Simon's position was revealed when I once interviewed him and asked him about the population crisis. He retorted: "What crisis?" and went on to say that there have never been as many people so well off and that there will never be a limit to population, because more people means more Einsteins to keep making life better. But neither Simon nor Perlman is a scientist.

If we inherit a bank account with a thousand dollars that earns 5 percent interest annually, we could withdraw fifty dollars or less each year forever. Suppose, however, we start to increase our withdrawals, say, up to sixty dollars, then seventy dollars, and more each year. For many years, the account would yield cash. But it would be foolish to conclude that we could keep drawing more from the account indefinitely. Yet that is what the Gees, Simons, and Perlmans believe. As the Atlantic groundfishery shows, we are using up the ecological capital of the planet (biodiversity, air, water, soil) rather than living off the interest. It is a dangerous deception to believe that the human-created artifice called economics can keep the indicators rising as the life-support systems of the planet continue to decline.

The value system that pervades most of the popular media not only perpetuates the delusion that resources and the economy are infinitely expandable, but also creates blinders that filter out the urgency and credibility of warnings that an environmental crisis confronts us.

Why a Warmer World Won't Be a Better World

IN AN ARTICLE IN *AMBIO*, THE NOBEL LAUREATE HENRY KENDALL and the population-dynamics expert David Pimentel point out that humanity is exceeding the Earth's capacity to support us all.

The community of Earth's diverse living things cleanse, alter, and regenerate air, water, and soil. Yet now we "either use, co-opt or destroy 40 percent of the estimated 100 billion tons of organic matter produced annually by the terrestrial ecosystem." In this way, we drive many other organisms that are keeping the planet habitable to extinction. Ozone depletion also has frightening repercussions, in that "of some 200 species studied, two-thirds show sensitivity to ozone damage."

And global warming, say the authors, "will be catastrophic for agriculture, changing rainfall patterns, drying some areas such as the central area of North America and increasing climatic variability. There will be effects on the growth of plants as well as collateral effects on plant pathogens and insect pests. Whole ecosystems may undergo major change."

This isn't just hypothetical or speculative. The hottest year on record so far, 1998, was "accompanied by a mid-continent drought which resulted in a 30 percent decrease in grain yield, dropping U.S. production below consumption for the first time in some 300 years. Similarly, Canadian production dropped about 37 percent."

Sustainable food production can be increased with better genetic strains and more efficient use of resources, but all have limits. To minimize global warming, fossil-fuel emissions must be cut drastically and forests must be preserved so that they can remove carbon dioxide. Improving life for all people while reducing emissions will require a staggering "three- or four-fold increase in effective energy services."

Kendall and Pimentel consider three possible scenarios in a warming world to the year 2050:

- Business as Usual (BAU)—Population rises to 10 billion; soil erosion, salinization, and waterlogging increase; no increase in aid for developing countries; no action to reduce global warming and ozone depletion.
- Pessimistic—Most dire predictions realized; high global temperature rise; high ultraviolet radiation; population growth to nearly 13 billion and worsening debt.
- Optimistic—Adoption of heroic efforts and major technological achievements; population stabilizes at 7.8 billion; energy-intensive agriculture is expanded; soil and water conservation improves; developed countries increase financial aid and technology; food is more equitably distributed and diets shift from animal to plant protein in developed nations.

The results of each projection:

- BAU—Grain land declines from 718 million hectares [277,000 square miles] in 1980 to 620 hectares [2.4 square miles] in 2050. There is 0.06 hectare [0.15 acre] per capita for grain production, less than a quarter of what was available in 1991. Food production in the developing world will be depressed by between 15 and 30 percent over the next twenty-five years; topsoil erosion will reduce rain-fed cropland by a daunting 29 percent. The result will be an average per capita loss in grain production in Africa, China, India, and other Asian nations of more than 25 percent.
- Pessimistic—This will lead to a further 15 percent decrease in grain production from BAU. Per capita production will be down by 40 percent despite a 30 percent rise above 1991 levels. For most of humanity, this means malnutrition and hunger.
- Optimistic—This scenario is based on almost doubling grain production by heroic efforts to increase irrigation and fertilizer use. Food production and environmental protection will have to be given the highest

> priority, and the cost and technological innovation would be carried by the industrialized world for the developing nations. If all people in the industrialized countries become vegetarians, food production is *tripled*, and the amount of energy expended to develop the world's agriculture is increased fifty to a hundred times, then, Kendall and Pimentel conclude, 7.8 billion people might be adequately fed by the middle of the twenty-first century. But they add with understatement, "This would appear to be unrealistic."

Kendall and Pimentel conclude with a stark warning and stiff challenge: "The human race now appears to be getting close to the limits of global food productive capacity based on present technologies. Substantial damage already has been done to the biological and physical systems that we depend on for food production.... A major reordering of world priorities is a prerequisite for meeting the problems we now face."

These are not the rantings of zealots announcing the end is nigh; they come from prominent scientists. Considering the media coverage given the death of Princess Diana and the sexual escapades of U.S. president Bill Clinton, surely Kendall and Pimentel deserve more attention than they've received.

SCIENCE AND ETHICS

IT IS IRONIC THAT MOST SCIENTISTS TODAY RECEIVE A DOCTORATE in philosophy without ever having taken a course in the discipline. As science exploded in the twentieth century, in order to stay abreast of discoveries and to acquire enough expertise to become a scientist, it became necessary to specialize earlier and in ever more restricted areas. When I graduated with a Ph.D. in 1961, it was possible to claim expertise as a geneticist and to be aware of virtually all aspects of heredity, from the molecular to the microbial, in plants and in animals, including humans. By the time I left the lab bench only two decades later, I identified my specialty as developmental genetics of the fruit fly, *Drosophila*.

Most of science continues to operate in the reductionist mode, focusing on a part of nature, isolating it, and controlling everything impinging on that fragment. And it has been a powerful way of knowing, providing insights that have liberated energy from atoms, taken us to the edges of the universe, and elucidated the entire sequence of the human genome. In our exuberance about our incredible discoveries, it is easy to overlook the immensity of our ignorance and to think that we are on the threshold of understanding virtually everything of importance. The elucidation of the sequence of DNA in the human genome was trumpeted as the Holy Grail of science that would generate cures for disease, explain human behavior, and enable us to control our destiny. Already scientists are backing away from those grand claims and are now hailing proteonomics, the interaction of proteins, as the real Holy Grail.

Many scientists themselves, caught up in the euphoria of discovery, have failed to recognize the weaknesses and limits of science. And certainly they are not trained or informed that there are responsibilities that accompany the practice of research, especially when carried out in universities with public funds.

Genetics after Auschwitz

THE BEST GUIDE WE HAVE TO HELP US THROUGH THE MAZE OF ethical questions that are created by genetic engineering is history; we forget its lesson at the risk of repeating the same mistakes. Consider Josef Mengele, the infamous doctor at the Nazi death camp at Auschwitz. He was in the news again a few years ago when forensic scientists eventually concluded that bones discovered in a grave in Brazil were indeed Mengele's remains. Recalling his activities should give every scientist pause.

Mengele gained his notoriety for his experiments in genetics. I was trained as a geneticist, yet never in all the years of my education or during my entire career as a scientist did I encounter his name except in the popular press. In the field of science, Mengele does not exist.

I went to a liberal arts undergraduate school that is ranked as one of the top in North America. My first genetics professor—my inspiration and hero—was a Jew who had received his Ph.D. from Curt Stern, one of the most important figures in classical genetics. Stern was also a Jew who had fled Nazi Germany and eventually ended up at Berkeley.

I earned a bachelor of arts degree, and though I was enrolled in honors biology, the curriculum stipulated that no more than half my courses be in science. So I had wonderful courses in music, Michelangelo, twentieth-century history, world religion, and literature. It was a marvelous opportunity rarely available to science students today, yet I never did encounter the philosophy or history of science. I was never taught about the excitement geneticists had felt at the turn of the century about the notion of improving the human species through selective breeding.

No one told us that geneticists had made bold claims about the "hereditary basis" for racial inferiority or superiority (which are value judgments, not terms that have any scientific meaning). Some of the leading geneticists in the twenties and thirties wrote of the genetic basis for nomadism in gypsies, criminality, drunkenness, and vagrancy, but we budding scientists did not learn of the important role that our predecessors had played in encouraging laws regulating immigration from certain countries, prohibiting interracial marriage, and sterilizing patients in mental institutions.

During my education in graduate school, I was never taught that being a scientist entailed enormous social responsibilities. We did not learn that the social context and value system within which scientific investigations occur affect the kind of research done and the way results are interpreted. No one told us that there are limits to science, that it provides only a fragmented view of nature that can never encompass the whole. There is no code of ethics governing our activity, nor is there a sense that it is a privilege for a scientist to have public support.

Today's science student has a heavy load of science courses and little opportunity to take others outside the discipline. Science students in my university don't have room among their courses for philosophy, history, religion, or literature. I remember one of the professors in my department during the student protests of the 1970s saying, "Okay, look. We'll keep all the best zoology courses for our honors and majors students and we'll *educate* the rest of them." Needless to say, he was sneering at the term *educate.*

Today, so much is happening in science that it is difficult for a scientist to stay abreast of all developments. Even though most research papers published will turn out to be wrong, trivial, or unimportant in a few years, we feel compelled to emphasize the latest work, thereby ignoring more and more of the classical studies and history. But what a loss that is.

We don't learn that geneticists were the prime movers behind the Nazi Race Purification program and that the voices of opposition to Hitler from scientists and doctors were silent. It was our colleagues, the likes of Mengele, who were carrying out their so-called research in death camps.

I know my fellow scientists find it easy to dismiss the doctors of death like

Mengele. We say, "Hell, he was a medical doctor carrying out pseudo-scientific experiments. He wasn't a geneticist at all, he was a nut." But "real" geneticists in Germany didn't say so. Besides, is it enough to write him off as a grotesque caricature of a scientist, a freak who happens once in a generation?

I don't think so. Germany boasted some of the most eminent geneticists and biologists of the day when the Nazi program was set in motion. The history of scientists not only in countries like Nazi Germany but in Fascist Italy, Japan, and during the McCarthy years in the United States is not a proud one. As Cornell University historian Joseph Haberer has written: "What becomes evident is that scientific leaders, when faced with a choice between the imperatives of conscience and power, nationalism and internationalism, justice and patriotism, invariably gravitated toward power, nationalism, and patriotism." It's a sobering indictment and one that can only be avoided in the future if we remember the past.

I do not believe for a minute that Josef Mengele was merely an aberration, who can therefore be lightly dismissed. Scientists are often driven, consumed, focused on the immediate problem at hand, and this is the great joy and strength of involvement in science. But it can also blind *any* of us to wider implications of what we and our peers are doing.

Mengele was one within a vast range of people who call themselves scientists. He was some scientist's student; he was a colleague, a peer of the medical-scientific community. It lets us off the hook too easily to say simply that he was a monster.

The explosion of the atomic bomb smashed the romantic notion of scientific innocence. The brouhaha over recombinant DNA and the vocal objections to the American Strategic Defense Initiative (Star Wars) are hopeful signs that the horrible silence of the scientific community may not happen again. But unless we acknowledge the likes of Josef Mengele and include him in courses taken by science students, we could quickly forget the lesson of history he provides.

THE RELUCTANCE OF SCIENTISTS TO TAKE AN UNFLINCHING LOOK AT their own history was clear when I was invited in the fall of 1987 to attend a meeting in Toronto of the organizing committee of the Couchiching Conference. Each year experts from various disciplines meet in

Couchiching, Ontario, to discuss a specific topic. In 1987 the subject was the Rise and Fall of the American Empire. In 1988 the conference was to look at DNA and genetic engineering, a timely topic in view of the tremendous advances in molecular genetics and the proposal to decipher the entire genetic blueprint of a human cell.

The Couchiching committee focused on the technology of DNA manipulation and its future implications, which are fascinating. But the really important questions have to do with what scientists and people in power will do with the unprecedented ability to alter the genetic makeup of life-forms. And the only way to anticipate that comes from looking backward.

At the Couchiching meeting, I reminded the committee that early in this century a brand-new science—genetics—had made spectacular and rapid discoveries about the laws of heredity. Scientists were understandably excited about the potential to apply this knowledge for the benefit of humankind. I pointed out that in prewar Germany, where culture and science were at a peak, doctors and scientists had embraced the possibility of applying the benefits of genetic discovery. By extrapolating from studies of the inheritance of *physical* characteristics in fruit flies and corn plants to *behavior* and *intelligence* in people, they concluded that human beings could be "perfected" through selective breeding and elimination of "defectives." The Nazi Race Purification programs seemed to represent the application of some of the most "progressive" ideas in science.

Thus, I suggested, doctors and scientists—especially geneticists—had, in their intoxication with new findings, popularized the notion of the overriding importance of heredity in human behavior and sold it to Hitler's National Socialists. This emphasis on heredity led inexorably to the horrors of the Holocaust, which scientists must therefore acknowledge some responsibility for. Two of the committee members (one a molecular biologist and both Jews) were outraged. They denied the suggestion that scientists have to bear some of the blame for the excesses of Nazi action and accused me of being "hysterical."

This selective memory of science's history amounts to a coverup and a revisionism that only ensures that the same thing could happen again. Even a suggestion that there is an unpleasant aspect to science's past is interpreted as opposition to science. I once hosted a television series that included a program presenting some history of genetics. In reviewing that show, Stephen

Strauss, the *Globe and Mail*'s science writer, wrote: "You are of the clergy, a scientist who left the monastery/laboratory to reform the world's understanding of his faith, and who now may well be on the way to becoming a heretic. (Some of your fellow geneticists think that.)" An accusation or even suggestion of heresy powerfully reinforces dogma and threatens dissenters with excommunication from the scientific community. I can understand why scientists are reluctant to face up to the past. Fortunately, a few exceptional ones won't let us forget.

BENNO MÜLLER-HILL IS A PROFESSOR OF MOLECULAR BIOLOGY IN Cologne, West Germany. Few texts in molecular biology and genetics fail to mention his work. While at Harvard University working with Walter Gilbert (who later earned a Nobel Prize), Müller-Hill carried out a classic experiment that allowed the isolation and purification of a protein molecule called a repressor, which controls gene activity. Only a few copies of the repressor are present in each cell, so the Müller-Hill/Gilbert experiment was a scientific tour de force. Müller-Hill's lab went on to collect sufficient quantities of the material to determine the primary structure of the protein. A scientist of world-class stature, he continues to do research. For the past decade, he has also studied the history of genetics in Nazi Germany.

One of Müller-Hill's articles, entitled "Genetics after Auschwitz," appeared in *Holocaust and Genocide Studies.* It is a document that is at once chilling and agonizing—an unflinching look by a scientist at the role of scientists in the Holocaust.

The article opens this way: "The past must be recollected and remembered before it can be evaluated.... It is particularly difficult for scientists. Science is oriented to the present ... only today's results exist. Only new data or new theories bring glory, honor and money for new research. Reflection on the past almost excludes the reflecting scientist from the ranks of present-day science." He goes on to summarize his studies and his book: "The rise of genetics is characterized by a gigantic process of repression of its history." Later he writes: "[G]eneticists have refused—and even now refuse—to acknowledge their history."

Science transcends national boundaries because it is practiced by an international community sharing knowledge through freely available publications. As governments around the world tie scientific innovation to their economic well-being, scientists are under tremendous pressure to do work that is socially "relevant" or "practical." Most research grants are now awarded and renewed on the basis of these considerations, and the potential economic rewards of an application of new ideas is great. Scientists focus even more than ever on work going on now and in the future. But if scientists, for the most humane of reasons, can participate in work that exceeds ethical lines, then surely it behooves us to pay attention to what has happened in the past. Today most scientists are too busy, or they ignore or selectively recall their history.

When Müller-Hill went to the archives of the German Research Association in 1981 to look at extensive historical records, he was told that he was the first person to do so since the end of the war. He was seeking answers to the questions, Was Auschwitz the result of pure scientific thinking? How does scientific reason change into the greatest unreason? Is this an inevitable process of the growth of pure scientific reason? Even asking such important questions carries the risk of denial and hostility from the scientific profession.

The German biographer P. Fischer illustrates this risk in his biography of Nobel laureate and geneticist Max Delbrück. In 1947, Nobel laureate H.J. Müller, as president of the American Society of Genetics, asked Delbrück, who was returning to his native Germany, "to gather first-hand information about geneticists still in Germany, and to investigate whether any evidence exists that might absolve them from the guilt of having actively supported Nazism and prostituted genetics under the Nazis." Delbrück failed to do so because "he did not have the courage; he considered it improper to inquire which German scientists had related to the National Socialist government. So he failed to fulfill the mission entrusted to him by the American Genetic Commission.... Later Delbrück was convinced by one of his closest friends, plant geneticist Georg Melchers, that virtually no German biologist had ever worked towards furthering the race theories, and that biology in general was not guilty of formulating the inhuman ideologies which resulted in inhuman action." Thus, even a scientist as great as Delbrück could not resist the pressures from his peer group and finally accepted the party line.

In the article "Genetics after Auschwitz" Müller-Hill says:

> The number of medical doctors who guided scientific selection in Auschwitz by killing through gas or slave labor was small. Nevertheless nine university professors were actively involved in selecting 70,000 mentally ill persons for the gas chambers.... The exact numbers of Jews killed in Auschwitz is unknown. Two-and-a-half million is the number Höss got from Eichmann.... Auschwitz was not only a site of destruction and a laboratory of human biology; it was also planned as a place of chemical production. IG-Farben had placed their largest investment during the war near Auschwitz.... When one regards the potential for scientific investigation, production and destruction, Auschwitz turns into a monument of modern science and technology. Human biology and technical chemistry should never have been the same again after Auschwitz was liberated, but nothing changed. The human biologists and medical doctors who were not caught in the act escaped by a semantic trick; this was pseudo-science or pseudo-medicine, they said, so they were free to start again with real science and medicine.

Müller-Hill traces the web of German scientists who were connected to the work carried out at Auschwitz. It included some of the leaders of the day. The most infamous of the scientists at Auschwitz was Josef Mengele.

Mengele was not just some scientist from the provinces. He had worked with the foremost German scientists of his time: first with the anthropologist Professor Theodor Mollisson in Munich, then with the human geneticist and specialist in internal medicine, Count Professor Otmar von Verschauer. Mengele was not an amateur who operated outside the established structures of science. Von Verschauer, Mengele's mentor, applied for grants for work at Auschwitz on "human twin studies, human eye defects, human tuberculosis and ... specific serum proteins." The grants were all approved.

Müller-Hill continues:

> What happened to the twins? They were analysed anthropologically, physiologically and psychologically. Anything that could be measured was measured. Those who had interesting anomalies

> were killed by Mengele or his helpers. The interesting organs were sent to the K.W.I. for Anthropology at Berlin-Dahlem....
>
> Was Professor von Verschauer a unique case? No, other researchers had already profited from the murder of the insane.... Some 70,000 patients of German mental institutions were killed by gas during 1940 and the first half of 1941.

Many of their brains ended up in scientific laboratories.

What is astonishing is that to most geneticists these infamous scientists and events don't exist: they have been expunged from history.

FOR BIOLOGISTS, NAZI GERMANY PROVIDES IMPORTANT LESSONS that might temper our rush to exploit new ideas uncritically. But according to Benno Müller-Hill, there has been a systematic suppression and revision of the history of science under the Nazis. Müller-Hill provides an explanation of how this came about:

> Auschwitz had just reached its highest destructive potential when the paper appeared which showed that DNA was the basic genetic material. It took several years until the significance of this discovery was generally understood, but when the double helix was published in 1953, only fools did not realize that genetics had virtually exploded. The speed of this development left no time for looking back or for regrets over the blood and tears that had been spilled in the process. Scientists discuss George Orwell's *Animal Farm* and *1984* and do not see that they themselves have created a universe which is equally frightening. No secret police forces them to forget the past. They obliterate it themselves on the marketplace of science. They have come to believe that they have a beautiful past, or perhaps no past at all. The chapters of textbooks which deal with genetics and society contain only a few token sentences about National Socialism.

If history is not remembered, can the scientific community be involved in horrors like those of Nazi Germany again? Of course, though undoubtedly in a different manifestation. Scientists are, above all else, human beings with all the foibles, idiosyncrasies, and diversity found in any other group of people. Ambition, driving curiosity, desire for power, thirst for financial security, fear—there are many reasons why people do what they do. And in the current scramble to capitalize on the enormous potential of genetic engineering, organ transplants, and a cure for AIDS, individual scientists have not been above cutting corners or compromising on ethical standards.

In part, the very methodology of science itself makes this easier, says Müller-Hill:

> Scientists observe and analyse objects. An object is a thing without rights. When a human being becomes an object he is nothing but a slave. What interests the scientist is the answer to the question he asks the object, but not the object's own questions. In general, the scientist never analyses the whole object but only a small part of it. Others dismember the object, he receives only one part of it for his analysis. The answers which he expects from the part he analyses may be numbers, DNA sequences or images.... This process of objectivization of the whole world, and finally of oneself as part of science, seems the main interest and pleasure of the scientist's brain. There is little place for other things in the scientist's mind.

But, it is often countered, weren't the people who carried out the atrocities in Nazi Germany second-rate intellectuals, mediocre but ambitious opportunists? Müller-Hill disagrees: "It was not in the interest of the Nazi elite that the sciences be dominated by a mob of liars and charlatans. The Nazis needed functioning science and technology to assist their wars of robbery and destruction."

But could there be a repetition of what was done by the Nazis?

> The killing of deficient newborn babies as practised in Germany between 1939 and 1945 has simply become anachronistic. Most geneticists sincerely believe that here they have created new values. They do not see that they appeal to the forces of the market which state that cost-efficiency considerations make it advisable, for both parents and state, to destroy the cost-inefficient embryo.

Müller-Hill's ideas are not pleasant, and he has encountered naked hostility from his scientific peers. But unless we hear him out and dig out the bad as well as the good in science's history, we will ensure that scientists will continue to do terrible things for what seem to be the highest reasons—just as their predecessors did. We need to remember that scientists are human too.

Geneticist R. Gold wrote a letter to the *Globe and Mail* to rebut the columns I wrote on Müller-Hill's work, yet Gold's letter itself provides an illustration of how revisionism is enforced by well-meaning scientists.

One tactic of the revisionist is to set up a straw man that can be knocked down. Thus, he writes:

> In his columns, Dr. Suzuki gives intermittent indications that he wants to go further than this to argue that there is a sort of evil inherent in the science of genetics and in geneticists themselves. He seems to imply, that if we are not carefully watched, we will be up to our old tricks again and indeed may already be engaging in nefarious schemes to harm humanity.

He goes on to say that I imply "that evil occurs because scientists are prone to evil."

He infers exactly the *opposite* to what I have written throughout my career in popularizing science. I have constantly emphasized that scientists are *no better or worse* than any other group of people, but it was enthusiasm about new insights in the mechanics of heredity early in this century that led some of the finest scientists to proclaim that human beings could be "improved" through selective breeding. The goals were laudable—to avoid suffering and improve the human condition—yet those ideas were warped into Nazi race purification, again, with the encouragement of some scientists. In our current excitement about the rapidly accumulating manipulative powers of DNA, many of the same claims are again being made.

Gold is a geneticist and surely knows that one of our colleagues at Harvard deliberately violated the federal guidelines on recombinant DNA experiments and, when discovered, quit the university to set up his own private biotechnology company. It was a UCLA doctor who violated federal grant restrictions by administering DNA to children with hereditary diseases. A Montana professor went ahead with prohibited field studies of DNA-engineered organisms in elm trees. There are other deliberate violations of regulations by reputable scientists. It is precisely because they are *not* evil or fanatics, but ordinary human beings who are totally caught up in their own ambitions and beliefs, that we must remember the pitfalls that the past reveals.

As I've mentioned, Müller-Hill pointed out another way that postwar German scientists rationalized what happened during the Holocaust so that the events could be discounted: "This was pseudo-science or pseudo-medicine, they said, so they were free to start again with real science and medicine." That's precisely what Gold does: "The concept of racial purity is not a scientific concept and has no place in genetics. It is simply intellectual rubbish that was dragged in to justify actions undertaken for other reasons." Thus Gold rewrites history.

Early in this century, leading scientists—geneticists and anthropologists—had popularized the idea of scientifically improving the human condition by preventing "inferior" people from reproducing or by encouraging those considered "superior" to have more children. They established the social climate for the adoption of race purification policies by Nazi Germany. Seen today, these ideas are indeed "intellectual rubbish," but in the twenties and thirties they were serious scientific proposals. The important lesson is that we should be very careful about rushing to apply in society or the ecosystem ideas derived in the laboratory. But we won't learn that essential lesson if we persist in papering over the past.

Gold's final accusation is that I have not addressed "what we should do about all this. Should we stop doing the research that will continue to provide us with these choices, and, if not, what advice has he as to the choices we should make?" As a once-practicing scientist, I continue to take enormous vicarious delight in the insights gained by scientists and have long written of the need for better support of good science in this country. But it would be foolhardy to suggest that there are no detrimental or unpredictable negative consequences of the application of science.

Gold seems unwilling to face up to the vast changes that have happened in science and its relationship to society in a few decades. In half a century, human population has more than *doubled,* while the per capita consumption of resources in the highly industrialized nations has increased many times more. In the same period, the scientific community has greatly expanded its numbers, just as the interval between discovery and application has decreased radically. Surely, then, there must be constant reassessment of the changing relationship between science and society, and the best guide that we have is history. Accusations of heresy, neo-Ludditism, or anti-intellectualism become powerful means of discouraging critical dissent or inquiry. In the long run, that does an enormous disservice to the public *and* to science itself.

The Final Dance on Racism's Grave

CLAIMS THAT RACIAL TRAITS HAVE A BIOLOGICAL BASIS DELIGHT bigots because they appear to support the contention that different races should "stay with their own kind." Racism is exacerbated by excessive claims about the importance of heredity on human behavior by geneticists. As the media headline sensational stories of racial conflict in our major cities, we should not forget that enormous social changes have occurred over the past decades.

I walked into my favorite bar a while ago and was hailed by a friend who is a regular there. As I ambled over, I could see that he was arguing with someone I didn't recognize. People in a circle around them were hushed and kept looking over at the arguers surreptitiously. As I approached, the stranger wheeled on me and demanded, "What do *you* people want?"

Looking back on that moment, I am amazed at the brain's demand for understanding. I hadn't heard a word of their argument and this question was totally incomprehensible, yet immediately my brain began to try to make sense out of it. Faster than you can read this line, my mind had decided that since my friend also worked at the CBC, the issue must have been the recent budget cuts.

"Look," I said, wading in, "there should have been an increase, not a cut."

"No, David," my friend interjected. "That's not what it's about." Instead of helping, that threw me back into confusion, and then my brain began a search down various alleys trying to make sense out of "you people."

My friend used to live in British Columbia; could they be discussing the province's economy? Or was it all of us who had gone to university? Perhaps it had to do with everyone who works out nearby at the YMCA. Gradually, my dim wits began to realize that he had lumped my friend and me together because although we are both Canadians by birth, he is of East Indian extraction and I am of Japanese descent. He meant all of us "colored" Canadians. What an irony when this guy turned out to have emigrated here from Central Europe when he was a child! Being an articulate academic, I turned on the full heat of my scorn and persuasively told him to "[expletive deleted] off!" and walked away. I showed him not to tangle with someone who has a way with words.

My point in relating this story, in addition to reflecting on how the brain functions to make sense of the world around it, is to share an experience that every member of a visible minority has had in Canada. These days, our media are full of stories of racial incidents. But I think they've got it all wrong. Sure, there is a lot of bigotry here—show me a country where there isn't. But don't let anyone tell you it's worse now than it was.

Each day that I take the bus to work in Toronto, I pass Jesse Ketchum School (it could be any other school in that city), and it fills me with delight to watch youngsters of every color playing, fighting, hugging—in short, doing what youngsters do but with total unawareness of visible differences. There is in Toronto today a mix of races that would have been impossible to foresee when I was a kid. Back then, most countries of origin of visible minorities were allowed to send one hundred immigrants a year to Canada.

This country is a bold experiment. Modern Canada is grounded in the historic rivalries between those of French and English extraction and the paternalistic oppression of the original inhabitants. There have been a lot of racial horror stories and mistakes and there will be more, but we have learned from past errors, and as we become aware of our often unconscious biases, we do change.

I think Canadians (and I'm one of the guilty ones) are too self-critical—we judge ourselves very harshly. But we shouldn't forget the pluses. My eldest daughter, a fourth-generation Canadian but genetically pure Japanese, never had a moment's concern that her husband is a Caucasian (albeit from Chile), yet in my parents' day, interracial marriage was unthinkable.

My parents, though born in Canada, could not vote until 1948, had to face a quota system in the universities, and could not buy property. Shortly after the war, my family moved to Leamington, a town in southern Ontario, where kids would boast that "no black has ever stayed in town past sunset."

My point is not to decry past injustices but to show that what was once considered OK is simply not acceptable today. Yes, there is still a lot of bigotry and we must always fight it when we see it, but I believe that for every discriminatory episode, there are dozens of acts of generosity, friendship, and assistance that cross racial barriers. They just never make the papers.

I was amazed and filled with pride at the response of Canadians to the Vietnamese boat people and to the starving Ethiopians. It was generous and genuine. Our record of assistance to the Third World, through agencies like CIDA (Canadian International Development Agency) and IDRC (International Development Research Centre), has been an example for the rest of the world.

And as we grope toward our ideal of equal opportunity and justice for all, we can look to those children in the school ground with hope. I'm reminded of two episodes in parochial Leamington of the 1940s that give me faith in youngsters. One of my great chums went to a club once a week where he had lots of fun playing games. He wanted me to join too, and I was delighted at the prospect. So I tagged along with him, but when I got there, the grown-ups made me sit outside the entire evening. My friend had wanted me to join the Sons of England!

The other experience involves another pal. We were playing together when my dad rode by on a bike. I waved and hollered at him, to the amazement of my friend. "How do you know him?" he asked.

"That's my dad," I replied.

"But he's a Chinaman!" he exclaimed. Yes, kids are color-blind.

Through Different Eyes

CHILDREN HAVE SUCH A DELICIOUS WAY OF IGNORING ALL SOCIAL conventions to ask the obvious question. That's what happened the other day: a child stopped on the street and said to me, "Hey, mister, how can you walk with your eyes closed?"

Evolution selected the epicanthic fold of eyelids characteristic of my Mongoloid ancestors. Those fat-filled lids worked well in frigid climates, insulating eyeballs against the cold while blocking out a lot of the reflected light from ice and snow. All you need is a slit to let light into the pupil—my field of vision is just as great as that of the "big eyes" of the West.

But what is functional and adaptive biologically is not necessarily acceptable culturally. Small eyes are not in vogue, and Western standards of beauty are pervasive. Even in Japan today, it's not uncommon to see people walking around with one eye bandaged because they've had a cosmetic operation to put a fold into the eyelid to give the illusion of bigger eyes. (They do one eye at a time.)

I'm not immune. In the years immediately following the incarceration of Japanese Canadians as enemy aliens, my eyes were a great source of shame to me. I would have given anything to have had an eye operation. (I wanted to dye my hair and change my name too.) I'm terribly self-conscious without my glasses, because I live with the illusion that the frames make my eyes look bigger. So there you have the primal essence of my hang-ups.

You may ask, "Who cares?" or "So what?" Well, I'm sure a lot of people have their own personal sensitivities to what they perceive as blemishes.

After each of my children was born and I was assured they were normal, the first things I checked out were their eyes and noses. Today, people deliberately limit the number of children they have, and hence each birth has become vested with far greater importance. We want so much for our children and would like them to start out with the maximum of advantages. So if music affects the fetus, we'll play Beethoven every chance we get. What will happen as techniques for monitoring our babies increase the kinds of traits that are detectable and alterable before birth? If a physical feature has come to be regarded as an impediment, imaginary or not, most of us would like to avoid it in our children, if possible.

Already the crudest kind of prenatal diagnosis by amniocentesis and chorion biopsy allows a quick detection of sex. In clinics in India and China, girls (who have an XX chromosome constitution) are not as desirable as boys (who have an XY chromosome makeup), so through abortion females are being "terminated." Their "defect" is the absence of a Y chromosome. Sex selection by abortion happens here too, though doctors are understandably reluctant to discuss it. Consider a family in which three or four girls have been born. The parents would like to have a son, and if the mother is in her late thirties, she is eligible to have amniocentesis (the recovery of fetal fluid and cells from the amniotic sac of a pregnant woman) to check for Down's syndrome in the fetus. After the test is done, parents can ask what the sex of the fetus is. If the fetus is a normal female, this family might contemplate an abortion in order to try for a boy. This does occur in North America.

The problem for society is that once the technologies become available, all kinds of unanticipated uses are thought up. Amniocentesis yields a crude biological profile of the fetus. Such information has allowed parents who carry genes for a known hereditary defect to risk pregnancy, knowing that a defect can be detected in the test and a decision can be made whether to abort. Access to the test has resulted in the birth of babies who otherwise would never have been deliberately conceived, and a great deal of worry and the burden of a defective child have been avoided. There are many terrible neurological and anatomical defects that can now be circumvented, but for many (including me) abortion of a Down's syndrome fetus takes us into a troubling "gray zone." I know families that have been devastated by the birth of a Down's child, but have also known others whose lives have been enriched in many ways by a Down's infant. Once we have options, we are faced with the terrible decision of where we draw the line and what we are willing to tolerate.

Now with ultrasonography, fetoscopy, chorion biopsy, and restriction enzyme mapping, techniques are rapidly evolving to provide very sophisticated insights into the genetic makeup of fetuses. And if Down's syndrome raises difficulties, what of those diseases in which sound minds are accompanied by progressively worsening physical conditions—muscular dystrophy, hemophilia, sickle cell anemia, cystic fibrosis, multiple sclerosis, and others? Would those who suffer from these debilitating diseases and their family members prefer that they had not been born? And as technology improves, what will we decide about scoliosis, diabetes, cleft lip and palate, albinism?

Already the U.S. government has given its approval to begin gene therapy, the deliberate introduction of DNA to alter the genetic makeup of children. Like the technologies of prenatal diagnosis, this will open up an awesome array of possibilities that many will find irresistible. I am glad that my reproductive days are over so that I don't have to deal with these very real possibilities. For if the option had been available, would I have been able to resist tampering with the conformation of my children's eyes?

The Temptation to Tamper

WHAT IS NORMAL? OR PERHAPS IT SHOULD BE ASKED, WHAT IS ABNORmal? That is the troubling issue raised every time scientists develop technologies that hold out the possibility of overcoming defects. For example, sterility in some men and women has existed since time immemorial. It is not a life-threatening condition, but when artificial insemination and embryo transplants became possible, sterility became a medical problem. A lot of medicine is now concerned with treating desires (smooth skin or large breasts), thereby converting normal human variability into abnormalities or defects.

What happens when we explore the possibilities that flow from genetic engineering? Through biotechnology, we can now engineer microorganisms to do our bidding. Many molecules normally made by the human body can now be manufactured in large quantities by bacteria.

One of the most spectacular feats in this area was the synthesis of the gene specifying human growth hormone (HGH). The gene was inserted into bacteria, which then turned out a protein identical to HGH. This hormone is normally produced in the pituitary gland, a small lobe of the brain.

HGH is an important growth regulator, and when it is missing or present in low amounts, pituitary dwarfism results. In the past, HGH was extracted from the pituitary glands of cadavers and used to stimulate growth in youngsters whose height at maturity was projected to be well below average. In the United States, the National Hormone and Pituitary Program processed up to 50,000 glands a year to supply 3,500 children and adolescents.

HGH was in limited supply and was expensive. This achievement has meant that there is a limitless supply of HGH and it should therefore become relatively cheap.

But now supply far exceeds demand. So in fine free-enterprise style, there has been a search for new potential markets for this product. Very little is known about the subtle biological effects of HGH. We know it is abnormally low in pituitary dwarfs, but we don't really know how well correlated the amount is with the height of the population generally. It does appear that HGH induces the loss of fat without a loss of muscle. On that basis, HGH is being proposed as a treatment for obesity. Meanwhile, athletes, ever watchful for any possible advantage in their competitive sports, are already taking the drug to induce preferential buildup of muscle mass. I can't overemphasize the extent of our ignorance about the normal metabolic role of HGH.

But the greatest potential to market plentiful HGH lies in the redefinition of what is abnormally short. Dwarfs lacking HGH are clearly defined. But in a distribution curve of normal people, there will always be those who fall at the short end of the height spectrum. At what point does a normally short person classify as "abnormal"? Apparently it's all in the mind. A UCLA professor described a father who brought in his son to be tested for HGH deficiency. When told that there was no cause for concern because all indications were that the boy would grow to a height of five feet seven inches, the father roared, "That's absolutely unacceptable!" and demanded HGH for his son.

Ours is a society in which size is connected to social rewards. Executives and bank presidents are taller than average, and bishops exceed priests in average height. Certainly in North America the rewards of size on the athletic field are staggering. The take-home pay of a towering basketball star or a football behemoth is immense. Any sports fan knows how fanatically some parents pursue a career for a child. It is not at all wild speculation to suppose that ambitious parents will line up to purchase HGH for their children. They will willingly put their children at unknown risk of side effects of the treatment. Doctors already report requests for HGH "therapy."

What we are talking about now are perfectly normal, healthy children who are going to be treated as if they are somehow defective and in need of medical treatment. A report last fall in the journal *Science* suggests that "it is almost certain that many affluent parents of short children will have their children treated with human growth hormone as a matter of course."

James Tanner of the University of London predicts that HGH treatment may become as accepted as orthodontics. Some scientists suggest that there is a need to do more experiments, to find out what the effects of HGH therapy might be. They speak of the need for placebos and double-blind experiments. But why? There ought to be a total ban on any use of the hormone on children who fall within the normal range of human variability. Medicine and science ought to be directed to more serious problems.

Update

In July 2003, a U.S. Food and Drug Agency (FDA) Advisory Committee recommended that the FDA approve pharmaceutical giant Eli Lilly and Company's application to market its synthetic human growth hormone known by its brand name, Humatrope, for a newly classified pediatric "indication." It is called non–growth hormone deficient short stature or NGHDSS, and what it refers to is children who are completely normal in terms of HGH but at the short end of height. Up to now, Humatrope has been used in Canada and the U.S. only for children with serious medical problems.

The FDA seldom ignores the advice of one of its own panels, so chances are very good this recommendation will be given the go-ahead. This means that up to 400,000 children in the U.S. could be eligible for Humatrope "treatment" because they fall within the clinical definition of short stature. It means that thousands of children could become part of a massive experiment as they begin to receive treatment to increase their stature by a few centimeters. Just as the pharmaceutical industry turned menopause into a medical condition to justify hormone replacement therapy, this will sentence perhaps one out of every hundred children to a new medical condition of NGHDSS. Eli Lilly spokesperson Morry Smulevitz argues that just because the biological cause of short stature in NGHDSS children is not known, "it is unfair to require that these children be denied therapy because a cause for their growth failure is not known." So far Eli Lilly and Company has not applied for similar approval for NGHDSS in Canada.

The Pain of Animals

MEDICAL TECHNOLOGY HAS TAKEN US BEYOND THE NORMAL BARRIERS of life and death and thereby created unprecedented choices in human lives. Until recently, we have taken for granted our right to use other species in any way we see fit. Food, clothing, muscle power have been a few of the benefits we've derived from this exploitation. This tradition has continued into scientific research, where animals are studied and "sacrificed" for human benefit. Now serious questions are being asked about our right to do this.

Modern biological research is based on a shared evolutionary history of organisms that enables us to extrapolate from one organism to another. Thus, most fundamental concepts in heredity were first shown in fruit flies, molecular genetics began using bacteria and viruses, and much of physiology and psychology has been based on studies in mice and rats. But today, as extinction rates have multiplied as a result of human activity, we have begun to ask what right we have to use all other animate forms simply to increase human knowledge or for profit or entertainment. Underlying the "animal rights" movement is the troubling question of where we fit in the rest of the natural world.

When I was young, one of my prized possessions was a BB gun. Dad taught me how to use it safely, and I spent many hours wandering through the woods in search of prey. It's not easy to get close enough to a wild animal to kill it with a BB gun, but I did hit a few pigeons and starlings. I ate everything I shot. Then as a teenager, I graduated to a .22 rifle, and with it I killed rabbits and even shot a pheasant once.

One year I saw an ad for a metal slingshot in a comic book. I ordered it, and when it arrived, I practiced for weeks shooting marbles at a target. I got to be a pretty good shot and decided to go after something live. Off I went to the woods and soon spotted a squirrel minding its own business doing whatever squirrels do. I gave chase and began peppering marbles at it until finally it jumped onto a tree, ran to the top, and found itself trapped. I kept blasting away and grazed it a couple of times, so it was only a matter of time before I would knock it down. Suddenly, the squirrel began to cry—a piercing shriek of terror and anguish. That animal's wail shook me to the core and I was overwhelmed with horror and shame at what I was doing—for no other reason than conceit about my prowess with a slingshot, I was going to *kill* another being. I threw away the slingshot and my guns and have never hunted again.

All my life, I have been an avid fisherman. Fish have always been the main source of protein in my family, and I have never considered fishing a sport. But there is no denying that it is exciting to reel in a struggling fish. We call it "playing" the fish, as if the wild animal's desperate struggle for survival is some kind of game.

I did "pleasure-fish" once while filming for a television report on the science of fly-fishing. We fished a famous trout stream in the Catskill Mountains of New York State, where all fish had to be caught and released. The fish I caught had mouths gouged and pocked by previous encounters with hooks. I found no pleasure in it, because to me fish are to be caught for consumption. Today, I continue to fish for food, but I do so with a profound awareness that I am a predator of animals possessing well-developed nervous systems that detect pain. Fishing and hunting have forced me to confront the way we exploit other animals.

I studied the genetics of fruit flies for twenty-five years and during that time probably raised and killed tens of millions of them without a thought. In the early seventies, my lab discovered a series of mutations affecting the behavior of flies, and this find led us into an investigation of nerves and muscles. I applied for and received research funds to study behavior in flies on the basis of the *similarity* of their neuromuscular systems to ours. In fact, psychologists and neurobiologists analyze behavior, physiology, and neuroanatomy of guinea pigs, rats, mice, and other animals as models for human behavior. So our nervous systems must closely resemble those of other mammals.

These personal anecdotes raise uncomfortable questions. What gives us the right to exploit other living organisms as we see fit? How do we know that these other creatures don't feel pain or anguish just as we do? Perhaps there's no problem with fruit flies, but where do we draw the line? I used to rationalize angling because fish are cold-blooded, as if warm-bloodedness indicates some kind of demarcation of brain development or greater sensitivity to pain. But anyone who has watched a fish's frantic fight to escape knows that it exhibits all the manifestations of pain and fear.

I've been thinking about these questions again after spending a weekend in the Queen Charlotte Islands watching gray whales close up. The majesty and freedom of these magnificent mammals contrasts strikingly with the appearance of whales imprisoned in aquariums. Currently, the Vancouver Public Aquarium is building a bigger pool for some of its whales. In a radio interview, an aquarium representative was asked whether even the biggest pool can be adequate for animals that normally have the entire ocean to rove. Part of her answer was that if we watched porpoises in the pool, we'd see that "they are quite happy."

That woman was projecting human perceptions and emotions onto the porpoises. Our ability to empathize with other people and living things is one of our endearing qualities. Just watch someone with a beloved pet, an avid gardener with plants, or for that matter, even an owner of a new car, and you will see how readily we can personalize and identify with another living organism or an object. But are we justified in our inferences about captive animals in their cages?

Most wild animals have evolved with a built-in need to move freely over vast distances, fly in the air, or swim through the ocean. Can a wild animal imprisoned in a small cage or pool, removed from its habitat and forced to conform to the impositions of our demands, ever be considered "happy"?

Animal rights activists are questioning our right to exploit animals, especially in scientific research. Scientists are understandably defensive, especially after labs have been broken into, experiments ruined, and animals "liberated." But just as I have had to question my hunting and fishing, scientists cannot avoid confronting the issues raised, especially in relation to our closest relatives, the primates.

People love to watch monkeys in a circus or zoo, and a great deal of the amusement comes from the recognition of ourselves in them. But our relationship with them is closer than just superficial similarities. When doctors at Loma Linda Hospital in California implanted the heart of a baboon into the chest of Baby Fae, they were exploiting our close *biological* relationship.

Any reports about experimentation with familiar mammals like cats and dogs are sure to raise alarm among the lay public. But the use of primates is most controversial. In September 1987, at the Wildlife Film Festival in Bath, England, I watched a film shot on December 7, 1986, by a group of animal liberationists who had broken into SEMA, a biomedical research facility in Maryland. It was such a horrifying document that many in the audience rushed out after a few minutes. There were many scenes that I could not watch. As the intruders entered the facility, the camera followed to peer past cage doors, opened to reveal the animals inside. I am not ashamed to admit that I wept as baby monkeys deprived of any contact with other animals seized the fingers of their liberators and clung to them as our babies would to us. Older animals cowered in their tiny prisons, shaking from fear at the sudden appearance of people.

The famous chimpanzee expert Jane Goodall also screened the same film and as a result asked for permission to visit the SEMA facility. This is what she saw:

> Room after room was lined with small, bare cages, stacked one above the other, in which monkeys circled round and round and chimpanzees sat huddled, far gone in depression and despair.
>
> Young chimpanzees, three or four years old, were crammed, two together into tiny cages measuring 57 cm by 57 cm [22 inches by 22 inches] and only 61 cm [2 feet] high. They could hardly turn around. Not yet part of any experiment, they had been confined in these cages for more than three months.
>
> The chimps had each other for comfort, but they would not

> remain together for long. Once they are infected, probably with hepatitis, they will be separated and placed in another cage. And there they will remain, living in conditions of severe sensory deprivation, for the next several years. During that time they will become insane.

Goodall's horror sprang from an intimate knowledge of chimpanzees in their native habitat. There, she has learned, chimps are nothing like the captive animals that we know. In the wild, they are highly social, requiring constant interaction and physical contact. They travel long distances, and they rest in soft beds they make in the trees. Laboratory cages do not provide the conditions needed to fulfill the needs of these social, emotional, and highly intelligent animals.

Ian Redmond gives us a way to understand the horror of what lab conditions do to chimps:

> Imagine locking a two- or three-year-old child in a metal box the size of an isolette—solid walls, floor and ceiling, and a glass door that clamps shut, blotting out most external sounds—and then leaving him or her for months, the only contact, apart from feeding, being when the door swings open and masked figures reach in and take samples of blood or tissue before shoving him back and clamping the door shut again. Over the past 10 years, 94 young chimps at SEMA have endured this procedure.

Chimpanzees, along with the gorilla, are our closest relatives, sharing 99 percent of our genes. And it's that biological proximity that makes them so useful for research—we can try out experiments, study infections, and test vaccines on them as models for people. And although there are only about forty thousand chimps left in the wild, compared with millions a few decades ago, the scientific demand for more has increased with the discovery of AIDS.

No chimpanzee has ever contracted AIDS, but the virus grows in them, so scientists argue that chimps will be invaluable for testing vaccines. On February 19, 1988, the National Institute of Health in the U.S. cosponsored a meeting to discuss the use of chimpanzees in research. Dr. Maurice Hilleman, director of the Merck Institute for Therapeutic Research, reported:

> We need more chimps.... The chimpanzee is certainly a threatened species and there have been bans on importing the animal into the United States and into other countries, even though ... the chimpanzee is considered to be an agricultural pest in many parts of the world where it exists. And secondly, it's being destroyed by virtue of environmental encroachment—that is, destroying the natural habitat. So these chimpanzees are being eliminated by virtue of their being an agricultural pest and by the fact that their habitat is being destroyed. So why not rescue them? The number of chimpanzees for AIDS research in the United States [is] somewhere in the hundreds and certainly, we need thousands.

Our capacity to rationalize our behavior and needs is remarkable. Chimpanzees have occupied their niche over tens of millennia of biological evolution. *We* are newcomers who have encroached on *their* territory, yet by defining them as pests we render them expendable. As Redmond says, "The fact that the chimpanzee is our nearest zoological relative makes it perhaps the unluckiest animal on earth, because what the kinship has come to mean is that we feel free to do most of the things to a chimp that we mercifully refrain from doing to each other."

And so the impending epidemic of AIDS confronts us not only with our inhumanity to each other but to other species.

Update

It has long been argued (usually by sport fishers) that fish do not feel pain. If the hook in the mouth of a fish caused pain, it is reasoned, the animal would not fight against the pressure but would come toward the fisher to reduce the pain. But any fisher knows that the most exciting part of "playing" a fish is when the animal tugs, thrashes, and leaps frantically to escape, behavior that sure suggests fear and discomfort.

A more powerful argument that animals don't feel pain is the suggestion that pain is a purely conscious experience, with heavy input from sensory and emotional components. By this reasoning, for a fish to feel pain, it must have consciousness. But, it is said, the brain of a fish is too primitive for such attributes.

On April 30, 2003, Dr. Lynne Sneddon of the Roslin Institute in Edinburgh published a paper in the prestigious scientific journal *Proceedings of the Royal Society of London.* In the article, she reported injecting bee venom or acetic acid into the lips of rainbow trout, the controls being fish injected with saline or simply handled manually. The venom- or acid-treated fish responded with displays of "profound behavioural and physiological changes" over a period of time. They took three times as long as the controls to begin feeding again and immediately after injection would rub their lips in gravel or against the walls of the tank. The fish displayed a " 'rocking' motion seen in stressed higher vertebrates like mammals."

Sneddon's group also found receptors in the head called polymodal nociceptors that are known to respond to stimuli that damage tissues. They conclude: "Our research suggests noxious stimulation in the rainbow trout has adverse behavioural and physiological effects. This fulfills the criteria for animal pain." Animal rights activists hailed the work, while sport fishers were nearly unanimously reluctant to accept it.

Are There No Limits?

BECAUSE MOST BIOLOGISTS AND MEDICAL SCIENTISTS CONTINUE TO operate on notions of reductionism—reducing parts of nature to their most elementary components—concepts and insights gained in the fragmented scientific way are often applied for limited purposes. Yet they may have profound moral and philosophical repercussions that ripple far beyond the restricted view of the scientist. Nowhere is this more apparent and difficult than in our ability to intervene in the fate of newborn infants.

In 1986, Baby Andrew, an eight-month-old boy in Ontario, was suffering from a rare form of blood cancer called myelomonocytic leukemia. His case focused interest on the question of where medical practices should stop.

During the 1950s, when I was a teenager, leukemia was a death sentence, and even today, all victims of the disease who are not treated soon die. But now the statistics on patients with its most common form, acute lymphoblastic leukemia, are astounding—with a regimen of radiation and chemotherapy, 90 percent have remissions, and more than half remain free of the disease five years later and are considered cured. Medical science richly deserves the accolades given it for many achievements of this century in improving the quality of our lives and the average life span we can expect.

Baby Andrew's form of leukemia is less tractable to treatment, though up to three-quarters of the children with it who are treated may go into temporary remission, with perhaps a third of remissions lasting beyond a year or two. But treatment of children younger than a year old is far more difficult. They are usually treated more aggressively, yet their survival prospects are very poor.

Untreated leukemia patients do not suffer intense pain or prolonged weakness—they usually die from an inability to fight off an infection or from bleeding in the brain. Treatment involves the use of chemicals that kill rapidly dividing cells. In this way, blood-producing bone marrow is wiped out in the hopes of destroying the cancerous cells. However, rapidly dividing normal cells, most notably those of the gastrointestinal tract, are also affected. Thus, chemotherapy is accompanied by intense nausea and vomiting. The treatment creates a discomfort far more excessive than that of the disease itself. It takes four to six weeks for a doctor to determine whether the patient is responding to treatment. Overall, the prognosis was that at the very best, Baby Andrew had a one-in-four chance of long-term survival.

I have deliberately stuck to the statistical and medical facts, but they hide the terrible human dilemma of parents faced with the reality of a sick child, literally deathly ill at the very onset of life. Baby Andrew's parents decided the odds of his survival after treatment simply weren't worth the certain suffering that would be produced by the treatment. They chose, therefore, to refuse medical intervention, thereby assuring his early death. Child welfare authorities felt otherwise and took the case to the courts, where they succeeded in forcing the medical treatment of Baby Andrew.

The high degree of success of childhood leukemia treatment today is built on generations of parents who opted to allow doctors to try anything in the minuscule hope of some medical miracle. Those dying children were guinea pigs whose parents willingly let them become ciphers in accumulated data. And many of those children, though already assured of dying early, had their last weeks and months made all the more cruelly painful by the experiments of medical scientists. As a parent, I hope I'll never be faced with that kind of terrible choice.

But the question still nags—how far do we go? I don't know the current success rate of organ transplants, for example, but the first ones all had to end in failure, however carefully experiments were first done with animals. Perhaps there is a quality of life gained for liver or heart recipients that justifies the operation in young children today. But what was the rationale for the experience of Baby Fae, the infant recipient of a baboon's heart? Her brief life was a grotesque media sensation that had no scientific or human justification. Will we soon transplant a miniaturized version of the Jarvik mechanical heart into younger and younger "patients"? Will we continue to create more children like David, who was imprisoned in a germ-free plastic bubble in the hope that there might be a breakthrough for dealing with his immune-deficient condition sometime in the future?

Somehow it seems to me that we have forgotten to ask a most important question—are there no limits? Has death become such an unacceptable prospect that we will pull out all stops to stave it off? In the name of medical "progress," do we demand that everyone submit to medical intervention on the chance that a statistically significant prolongation of life may ensue? The imperative to fight death at all costs cannot stem from some profound commitment to the sanctity of life. If it did, doctors could not possibly support the current levels of abortion in major cities that terminate up to half of all pregnancies for reasons that have nothing to do with the health of the woman or the fetus.

In an increasingly secular age, science seems to have cut us loose from any sense of place and meaning. Cosmologists inform us that our Sun is a very ordinary star among billions in the Milky Way Galaxy which in turn is just one galaxy among billions. Biological science indicates that life arose on this planet by chance, that we have the form and shape that we do not from a divine creation but because of the vagaries of environment and natural selection. Bereft of meaning and alone in the universe, we find death all the more terrifying. So we choose to fight it, and each time death occurs we must admit to defeat by the forces of nature. But in fostering the notion that through science and research human beings can leap the boundaries of nature and free ourselves from the dictates of our own biology, we have created something even more frightening than death—massive medical intervention to prolong the process of dying. And it is an innocent infant with no fear of death who makes us face up to this terrible reality.

At the other end of the cycle of life, medical science and technology have derived powerful tools to intervene in the dying process, thereby rendering a definition of death much more difficult. Prolongation of life for its own sake has thrust us into the uncomfortable position of trying to play God.

On December 1, 1986, my twin sister's eldest child, Janice, fell into a deep coma from which she never regained consciousness. She was twenty-seven years old. Viral encephalitis swept through her brain, causing massive and irreversible damage. For four months, my sister and her family sat in vigil as medicine's arsenal of tools—tranquilizers to reduce the convulsions, antibiotics to limit bronchial infection, intravenous feeds to sustain nutrition and fluids—kept Janice "alive." I was astounded by the tenacity of the evolutionary mechanisms operating in Janice's body to maintain life.

Three years earlier, returning home from a movie with my father, my mother suffered a massive heart attack. She was seventy-four and for years had been showing the progressive memory loss of Alzheimer's disease. Fifteen minutes after she collapsed, she was resuscitated by paramedics, rushed to a hospital, and put on a respirator in intensive care. Oxygen deprivation had permanently damaged her already deteriorating brain, but as with my niece, all Mom's survival mechanisms had kicked into action and she "lived" for another week.

It is one of the ironies of the twentieth century that the great success of modern medicine has made death so much more ambiguous and, in many cases, agonizing for the patient and family. A century ago, my mother and niece would have had rapid and humane deaths. Don't get me wrong. I wish my mother and niece were still alive and healthy, and I am grateful for the emergency centers and extraordinary techniques and expertise now available in crisis circumstances. But is it monstrous to hope for death when the medically managed process of dying seems driven only to stave off at all costs an inevitable death? Often those costs are increased pain without a corresponding improvement in the *quality* of life that is gained.

We should not forget that no doctor has ever "conquered" death; at best, death is postponed. Like all other life-forms, each of us will die, and no amount of scientific research and heroic measures will ever change that. What we are rapidly accumulating is an ability to intervene and disrupt the natural course of events that (still) lead to unavoidable death.

Today, medicine is a costly and complicated business. We have dramatic surgical techniques, diagnostic machines, and an array of powerful drugs. But in a society in which nutrition and sanitation have virtually wiped out major killers of the past, new technology itself drives doctors to the treatment of ever more rare and exotic conditions. Whereas premature babies weighing less than 2 kilograms (4 pounds) once had a poor chance for survival, today they routinely make it. But that means that a 1-kilogram (2-pound) baby becomes a potential target for intervention. Once-lethal congenital defects are being corrected by radical surgery on newborn infants and even, in some cases, on fetuses. At the other end of life, has death become so terrifying that we'd prefer to be tethered by a 2-meter (6-foot) tube to a machine that keeps a mechanical heart pumping than to die of heart disease?

Once medical science has made new techniques available, we can't deny or revoke them. But when I read heart-rending stories about children who die before they can receive a third or fourth liver transplant, I feel relief for the child.

By intervening in the process of death, we create novel situations where there are often no biological models to guide us—essentially, doctors create a new kind of human being. A simple example is the development of a shunt to allow the draining of the fluid that often builds up in the skulls of children with spina bifida. Developed by an engineer who was horrified by the cranial swelling in his spina bifida child, that invention eliminated much of the brain damage and the disfiguring buildup of fluid in the skull. Children who before the shunt almost always died now survived.

Only then was it found that spina bifida is not a simple condition. The prognosis for life with the shunt varied from child to child. Based on experience, doctors found that the efficacy of installing the shunt depended on the spinal lesion's position, size, and severity. So today, in many British hospitals, doctors assess spina bifida infants at birth and place them in different categories. Some receive the shunt and every effort to provide a high quality of life, whereas others are allowed to die. It was technology itself that created this sorrowful dilemma.

There are many who do not want to face up to the decisions that now have to be made. Aiming for life at all costs, regardless of its quality, simply does not take into account the reality of today's technologically sophisticated world where issues are far more profound and difficult than they once were.

But medical doctors and all those who have lived with the agony of seeing loved ones suspended in that technological twilight between life and death cannot avoid the issues.

A BIOCENTRIC VIEW

INCREASINGLY, URBAN DWELLERS WHO SPEND THEIR TIME CAUGHT up in the frenzy of consumption and entertainment in cities have fewer opportunities to encounter the natural world. Even our parks have become places where it is possible to take all the urban amenities of television, refrigerators, and small vehicles with us while experiencing the wild.

So it's not surprising that one of the most common questions I'm asked is "Who needs nature?" It is not meant to be argumentative or confrontational but is an innocent query. The idea is that urban life is sufficient. In a human-created environment where we are surrounded by others of our species, our inventions, a few domesticated plants and animals, and a number of pests, it is easy to assume that our intelligence has enabled us to escape the limits of nature. Nothing could be further from the truth or as dangerous a conceit.

Why the Bravest Position Is Biocentrism

WE LEARN TO SEE THE WORLD THROUGH THE LENSES OF THE INDIVIDUAL beliefs and values that we have acquired from personal experience, family, and society. People often share a commonality of truths and values that are so widely accepted that they are seldom questioned.

We can't help seeing our surroundings through the perceptual filters of our preconceptions, yet the media continue to hold out an ideal of journalism that is "objective" and "balanced." But any journalist's personal values are bound to influence the "facts" selected and the way they are juxtaposed and arranged to create a story. The best way to strive for balance is to have many journalists presenting a wide array of worldviews.

Understandably, people in the media are preoccupied with human affairs—wars, budget deficits, sports, and entertainment. Even environmental stories are usually built around the human costs and benefits for health, esthetics, jobs, or the economy.

When wilderness habitats are invaded and species threatened with extinction, their preservation is often justified by their potential utility for human beings. Thus, it is pointed out that perhaps a quarter of the active ingredients of all medicines are natural compounds extracted from living organisms. When species disappear, a vast repertoire of potentially useful materials is also lost. It is also argued that wilderness may generate revenues through ecotourism or provide spiritual solace.

Everything we use in our homes and workplaces—electricity, metal, wood, plastic, food—comes from the Earth. Our economic system is based on our need for them and their scarcity or abundance. Consequently, the future of old-growth forests, coral reefs, or watersheds often rests on the merits of economic returns from protection or exploitation.

An "anthropocentric" ecological ethic recognizes that environmental protection is ultimately in our self-interest because, as biological beings, we depend on the integrity of our surroundings for our survival.

There is an alternative perspective called biocentrism, which Bill Devall, coauthor of *Deep Ecology,* defines as "a worldview emphasizing that Nature has intrinsic value, that is, value for itself rather than only aesthetic, commodity or recreational value for humans; that humans have the capacity for broader identification with Nature as part of our ecological self; and that compassionate understanding is the basis for communication with Nature as well as with other human beings." This is the central belief that underlies deep ecology.

Critics often accuse deep ecologists of being misanthropes, caring more for other species than for our own fellow human beings. I've heard it said derisively, "They want to protect trees and the spotted owl and don't care if people are thrown out of work." To such criticism, the U.S. poet Gary Snyder responds, "A properly radical environmentalist position is in no way anti-human. We grasp the pain of the human condition in its full complexity, and add the awareness of how desperately endangered certain key species and habitats have become.... The critical argument now within environmental circles is between those who operate from a human-centered resource management mentality and those whose values reflect an awareness of the integrity of the whole of nature. The latter position, that of deep ecology, is politically livelier, more courageous, more convivial, riskier, and more scientific."

When we acknowledge our dependence on the same biophysical factors that support all other life-forms, the responsibility for "managing" all of it becomes a terrible burden. In fact, it's an impossible task because, in spite of the impressive sophistication and progress in science and technology, we have nowhere near enough information to understand, let alone predict and control, the behavior of complex systems like watersheds, forests, oceans, or the atmosphere.

A Biocentric View

Amid the barrage of information from the print and electronic media, we must recognize the inherent biases that often flow from our anthropocentrism. For example, in all of the discussion about the catastrophic loss of northern cod or the fate of the old-growth forest of Clayoquot Sound, all of the "stakeholders" in the fishing, logging, tourism, and Native communities seem to accept that the underlying economic and political institutions are beyond question or change, even though they may well be the very cause of the crisis.

By looking at the world through biocentric lenses, we may recognize the roots of our destructive path. The landscape may be uncomfortable and strange, but we can't afford to dismiss the problems viewed from this perspective as a lack of balance or simple bias.

Borrowing from Children

ENVIRONMENTAL AWARENESS CAN BE DRAMATICALLY CATALYZED BY AN *issue that becomes a symbol.* DDT, PCB, *and* CFC *now instantly summon up the "costs" associated with human-made chemicals. Bhopal, Chernobyl, and* Valdez *illustrated the predictable occurrence of human error with all technologies. Whooping cranes, whales, and baby seals raised the question of the innate value and rights of other life-forms.*

And increasingly, the fate of entire ecosystems—the Franklin River in Tasmania, Antarctica, and the Amazon rain forest—have become contentious because they are the focus of opposing forces of preservation and development. In Canada, the Windy Bay watershed on Lyell Island in the southern part of the Queen Charlotte Islands off the northwestern coast of British Columbia focused efforts by the forest industry, Native people, and environmentalists for fifteen years. On March 21, 1987, I wrote the following column:

"We no longer inherit the earth from our parents, we borrow it from our children." This is the stark message on a Green Party poster, a warning well worth some thought. Most of us hope that our children will be better off than we are, that they will have lives qualitatively richer than ours. But now it's not possible to cling to that expectation. Each day, the planet becomes poorer, as species of animals and plants disappear forever and the vast pool of biological diversity shrinks.

Even as recently as the early part of this century, it took two men several days to cut down one of British Columbia's immense trees. Today, one man and a chain saw can do it in minutes. But we haven't compensated for the environmental impact of our numbers and technological power by scaling down the way we attack the Earth.

The first European settlers came to a continent teeming with life, but the way we exploited it is a terrible chapter in our species' history. Passenger pigeons were once abundant beyond belief, literally darkening the skies for days on end when they migrated, yet we exterminated them within a century. The great herds of bison suffered a similar fate. How much poorer we are for their loss, to say nothing of the loss to the animals themselves.

Don't we learn from history? Do we not have any sense of our place in nature that should limit our predation? We have always taken pride in the fact that our brains enable us to see beauty in the world and even to add to it through poetry and art. Does it not diminish us when we cause the extinction of other life-forms that we appreciate as unique and beautiful?

One of the most unusual ecosystems in all the world is British Columbia's Queen Charlotte Islands. Because some of the islands were not covered by ice during the last glaciation, plants and animals survived there that are found nowhere else in the world. And for thousands of years these islands, the forests, and the surrounding waters supported a rich and diverse civilization whose descendants are the Haida people.

Much of the northern part of the Charlottes has been denuded by logging, but the southern 15 percent, called the South Moresby Islands, is still relatively untouched. In this area is Lyell Island, which harbors a 2,800-hectare (6,900-acre) coastal rain forest called Windy Bay. It took thousands of years of plant succession and selection for the immense trees that now populate the watershed to grow. Some of them are more than a thousand years old.

For more than twelve years, environmentalists and the Haida have fought to protect the South Moresby region from logging by Western Forest Products, which holds the tree farm license in the area. Windy Bay has been the focus of the debate. During this entire contentious period, logging has been allowed to continue on the island around Windy Bay.

The British Columbia government's tactic has been to approve logging permits while its own public inquiry was going on. The first study group lasted three years and recommended the establishment of a more broadly based committee. So the South Moresby Resource Planning Team was set up, with representatives of logging, Haida, and environmentalist and government groups. It deliberated for five more years and recommended the preservation of Windy Bay. This report was put on the shelf and ignored for two years, and all the while, logging continued.

Under public pressure, the B.C. government again set up a study group in 1985, this time called the Wilderness Advisory Committee, and charged it with looking at *sixteen* disputed areas in the province. Furthermore, the government asked for recommendations on all of them within three months! Shortly after the committee's hurried report was turned in, a new government was elected and declared itself unbound by any previous studies. Logging has not stopped during the entire period.

The issue involved is not whether there should be logging anywhere—no one disputes the economic importance of logging. But the question is whether a few special wilderness areas ought to be left untouched. There are strong philosophical, esthetic, and practical reasons for leaving some remnants of nature as they once were.

It is the height of hypocrisy and cynicism to allow logging to continue in the special areas while public debate about their fate is still going on. The tactic is transparent—once the trees are cut down, they are gone forever and there is no debate. Those who maintain that they can replace the likes of what we "harvest" and that they can "manage" our resources and sustain the yield cannot see beyond immediate profit and will not admit to enormous ignorance. Future generations will judge them harshly.

Only our present generation, through strong and sustained pressure on government leaders, can hope to preserve a remnant of what was once the great Canadian heritage.

Under intense pressure from the public, both the federal and provincial governments negotiated furiously. For British Columbia, the fate of South Moresby became the symbol of the future path for loggers, environmentalists, and Natives. I wrote this column on June 6, 1987:

By the time this column appears, the future of the South Moresby area of the Queen Charlotte Islands may have been decided. Federal and provincial negotiators have agreed on setting the area aside as a national park. The final formal agreement will be a major achievement by environmentalists, and all Canadians should congratulate Ottawa and British Columbia for the successful conclusion.

The fate of the area has come down to a matter of money. British Columbia's government says about CDN $100 million should be set aside for the park, while the federal ministry of the Environment wants a spending limit near CDN $40 million.

This is in the tradition of the political game of poker. In all the discussions, however, two issues have been omitted that cannot be ignored in future deliberations over wilderness areas: Are the only things of worth those that are measured according to their use for humans? Are the only meaningful values those that can be expressed in dollars and cents? As long as we assess everything as a potential resource, then it's just a matter of thinking up a use and costing it out to decide on the fate of anything.

In January 1987, I attended a meeting in New Zealand at which I heard a bioethicist justify research on human embryos because it would eventually "allow us to recover the many embryos that are normally aborted and so are an enormous waste." This "resourcist" attitude reminded me of a chemistry class in high school in which the teacher informed us that if a human body were reduced to its basic elements—calcium, carbon, sodium, and others—those chemicals would be worth only a few dollars. (Perhaps on today's market a couple of hundred.) What impressed me at the time was the thought that those few bucks' worth of atoms when arranged as a living person had a value beyond measure. The teacher had given us a nonsensical way of looking at a human being.

Now, however, I realize that the teacher was on to something. As we seek ways to stimulate the economy, we are missing out on a valuable "resource"—human bodies. Today, we have sophisticated extraction procedures, and if we were to purify from our bodies complex components in the form of enzymes, lipids, sugars, nucleic acids, and so on, I'm sure we could reap thousands of dollars' worth of molecules. We might even discover new compounds that will fight cancer, help us lose weight, improve memory, or postpone aging.

We also can easily keep a body on life-support systems and thus increase its shelf life. So we could continue to harvest blood, hair, antibodies, skin, and other useful material for months or even years, increasing the body's value enormously. For that matter, a body kept in this way could be a valuable test system. We could investigate cancer-causing agents or the harmful effects of cosmetics. Uteruses could be used as receptacles to study the early stages of embryonic development, and males would yield a steady supply of sperm for artificial insemination.

Of course, the body is a gold mine as the source of organs for transplantation, because, to the recipient, an organ is a precious gift. How precious? Is a kidney worth $20,000? A kidney recipient who no longer requires expensive dialysis can resume a place in the workforce and pay taxes, so he or she won't be a drain on the economy. Clearly, the government should be involved in salvaging organs because the economic returns would be enormous. Obviously, the organs and tissues from a single human body could net a return of millions of dollars.

This perverse line of thinking could be extended to determining the best age at which to "harvest" a body, all part of a matter of dollars and cents. However, nothing in such an economic equation would encompass the worth of those organ systems as a whole, namely, in a living, functioning human being. A human being has a value beyond anything economic because of factors that simply lie outside monetary considerations.

And is it any different when we try to cost out the "value" of wilderness? The Windy Bay watershed on Lyell Island, for example, is pristine, a remnant of what there once was all along the west coast of British Columbia. It is one of the last of its kind, and when it is gone it will be gone forever.

Can we calculate its value other than through its potential for logging, tourism, and fishing? Environmentalists and the Haida know that there is a worth beyond measure, but as long as they must continue to counter the exploiters with "realistic" economic arguments, we lose sight of the central issue. And we will continue to harvest and destroy because of nature's "resource value." We have to raise the level of discussion of wilderness above jobs and economic growth.

As negotiations between federal and provincial governments went on, public pressure escalated, but I had to leave Canada on a five-week trip to film in the Soviet Union as the discussions about Windy Bay continued. While I was in Irkutsk,

Siberia, I received a desperate call from my wife informing me that the talks had fallen through. The B.C. government had cut off negotiations and decided to make a provincial park, thereby allowing logging and mining in the contended area. Frantically, I wrote the following column, which was then carried by hand to Moscow by one of the members of our crew, and from there it was telexed to the Globe and Mail *in Toronto:*

Pain. Grief. Anger. These are my feelings as I write this on the shores of Lake Baikal in Siberia, where Canada seems impossibly remote. I'm trying to make sense of the phone call I received in Moscow three days ago (June 17) from my wife. She told me that British Columbia premier Bill Vander Zalm has scuttled negotiations with federal environment minister Tom McMillan to create a national park of the South Moresby area in the bottom 15 percent of the Queen Charlotte Islands. After a decade and a half of rancorous and divisive debate and confrontation between loggers, Haida, environmentalists, and government, once again the value of the area has come down to jobs and logging revenue.

Poll after poll has revealed that Canadians list environmental issues as a top priority, one they are willing to pay for in taxes and jobs. During his term in office, former premier Bill Bennett's own pollsters discovered that an overwhelming majority of British Columbians favored keeping South Moresby as a park.

Federal environment minister Tom McMillan has gone a long way with the generous package presented to British Columbia. His final offer was almost three times as much as he told me he had available last April. He has made a heroic effort to preserve the area.

But the issue ultimately is not over money; it is a question of what our society values. Wilderness everywhere is disappearing in response to the global demand for more economic growth and more consumer products. We now point with alarm to other parts of the world at the vanishing tropical rain forests, species extinction, and habitat destruction through development. But how can we be so self-righteous when our own actions don't match our words?

There is an unseemly haste to start those chain saws again and to get into the Windy Bay watershed, the jewel in the crown of South Moresby and the thorn in the side of the loggers. Already, Frank Bebbin, the owner of the company logging the area, directed his men, in one of the most cynical moves, to clear the hills flanking Windy Bay while negotiations about the fate of the area were going on. As soon as the loggers go over the top, they will begin the rapid irreversible destruction of a pristine watershed. It will not be done without triggering a massive response from Natives and environmentalists.

The number of jobs being "threatened" by the park is fewer than seventy! The federal government has offered a generous package that would more than compensate those loggers for the loss of their jobs. Yet that handful of loggers—minuscule in comparison with layoffs in most other industries—has been able to marshal enormous muscle and outweigh the wishes of the majority of people in the province and the country.

What can we surmise from this? Several things:

1) Frank Bebbin and the logging industry clearly have a powerful "in" with the provincial government. Since Premier Vander Zalm's decision flies in the face of public support, we have to ask what gives a small logging company such political clout.
2) This issue must be ideological. If the park were approved, that would be perceived by the logging industry as a "victory" by the Native people and environmentalists, rather than as a wise decision made for all people. The B.C. government has thrived on confrontation politics, creating the fiction that preserving South Moresby represents a death threat to the logging industry and fostering the illusion that environmentalists and Native people must be enemies of loggers and millworkers.
3) At the heart of the issue is a conflict in "worldviews." One worldview looks at all of nature as a "resource," so that by leaving a "decadent" forest such as the one around Windy Bay, we "waste" it. The other sees wilderness as an immensely complex and balanced ecosystem whose components will never be fully comprehended but whose very existence enriches our lives spiritually. Whatever one's worldview, it is a fact that an ecosystem like the one in Windy Bay is not like Mr. Vander Zalm's Fantasy Garden (a reference to Vander Zalm's private park, which is run by the premier's wife, Lillian), it is the product of thou-

sands of years of evolution and natural selection and can never be duplicated by human beings.

Why is it that virtually every B.C. proposal for federal assistance includes a request for tens of millions for "reforestation"? The answer is that the logging industry's claims that it is "sustaining the yield" of a "renewable resource" is a cruel hoax. Globally, logging industries are plundering the planet and failing miserably to replenish the forests. That is what makes the remaining virgin stands so attractive—they represent much better profits than the so-called crops that the industry has sown.

Once trees like those in Windy Bay are cut down, there is no issue. But until the saws and axes touch those trees, there is time. The federal government has acted with courage and generosity, and all of us who care about showing some humility and leaving something for our children must now register our support for its initiative—massively and unequivocally. Write, telegraph, or phone the prime minister, minister of the environment, and your local MP. Register your disapproval to B.C. premier Bill Vander Zalm and his minister of the environment. Give money to groups fighting to save South Moresby and be prepared to put your body on the line.

It's time for people to stand up and be heard. Pierre Berton, Farley Mowat, the Nylons, Bob Bateman, you are friends of South Moresby. Speak up now. And what about the B.C. NDP? Where is Mike Harcourt, and what is the position of his party on logging South Moresby? What about the federal NDP and Ed Broadbent? John Turner represents a B.C. riding—what is his position and that of his party? This could be a pivotal issue in a growing political consciousness of the true value of wilderness.

The above column was never printed. With stunning speed, a massive public outcry arose after Vander Zalm's announcement. It forced a renewal of negotiations, and a brand-new federal park reserve was created. The outcome was overwhelming. It was an all-too-unusual victory for Native people and environmentalists alike, so I was able to write a rare celebratory column:

It is a moment for all people to savor and for Canadians especially to be proud: the South Moresby area of the Queen Charlotte Islands will be established as a national park reserve to be preserved in perpetuity. This is a splendid achievement with profound implications.

First, we have to congratulate B.C. premier William Vander Zalm for making a tough and wise decision in agreeing to the park. He negotiated a generous price from the federal government, for which all British Columbians ought to be grateful. And he made a clever strategic move with his initial announcement last month that the province would establish a park within which logging would continue. The ensuing massive response gave him the unequivocal public support for the park that he needed to back his decision to accept the federal offer.

Prime Minister Brian Mulroney's strong support for South Moresby will endure as one of his best political stands. There is no question that without his approval the park would not have come about. And last, but far from least, environment minister Tom McMillan must be recognized for his role in creating the park. When I spoke to him last May, he indicated that the most money he could scrounge from a cost-conscious government was about CDN $40 million. Somehow he found the support within Cabinet to increase the offer almost threefold. His refusal to compromise with British Columbia on the park boundaries was vital. At some point, politicians have to take stands on issues that matter more than getting reelected. Tom McMillan did that.

There are others who should be heroic role models for our youngsters. They are people who gave up careers, steady incomes, even relationships to devote a major chunk of their lives to the preservation of South Moresby. Each one deserves an article, but here I can only list them and hope they will not be forgotten: Thom Henley, John Broadhead, Colleen McCrory, Paul George, Bristol Foster, Vicki Husband, Tara Cullis, and Jeff Gibbs.

Artist Bill Reid was a spiritual inspiration for all, and the Haida people, led by Miles Richardson, put their culture and land claims on the line and even risked jail to save the area. Through their dedication, these people have left a priceless legacy to last beyond their lives. And I think Canadians should look in the mirror and congratulate themselves because it was their support that gave politicians the mandate they needed.

What have we learned from all this? South Moresby, especially Windy Bay, became a symbol for all special wilderness areas. It became a crucial issue because the outcome would indicate what Canadians value and how we see ourselves in relation to nature. The very act of debating the importance of wilderness, logging, tourism, fishing, Native rights, and mining was part of an educational process going on around the world.

I've just returned from a six-week filming trip in the Soviet Union. The film crew spent eight days at Lake Baikal, Siberia's magnificent treasure that contains as much fresh water as all our Great Lakes combined. For twenty years, Soviet citizens and scientists have been fighting to protect the purity of Baikal. With the help of *glasnost,* they appear to have won.

Logging around the lake has been stopped, a huge girdling park zone established, and the paper mill on the lake will be phased out within three years. Earlier this year, after much public pressure, Soviet leader Mikhail Gorbachev finally cancelled a megaproject that had been planned for years. It was a proposal to divert to the south several huge rivers that flow into the Arctic Ocean. It would have caused climatic changes, to say nothing of a massive ecological disruption. (The scheme was as crazy as the idea of converting James Bay into a lake for export of fresh water.) So even in a society as rigid as that in the Soviet Union, environmental values have become an inescapable public issue.

South Moresby does not represent the last skirmish or area to be fought for. Canada's total parks system is very small and must be increased so that we can establish a fragile network of wilderness, tiny oases of diversity that will be preserved as a hedge against our ignorance and as treasures for our children. We have to invest far more money in silviculture and reforestation. We have to study and copy other countries, such as those of Scandinavia, that do sustain their logging yields, because their forest resources are far more limited than ours.

In the end, what South Moresby revealed was a profound clash between worldviews. The dominant one sees all nature as a potential resource, of value only for its economic worth. But there is growing support for a different outlook that recognizes that we are biological beings who, in spite of science and technology, remain attached to and dependent on nature. We have to fight to keep nature intact and to try to bring ourselves into a balance with the environment. South Moresby could be a watershed that marks a shift toward this re-emerging worldview.

Making Waves

THE MOST FREQUENT CHALLENGE I ENCOUNTER AS AN ENVIRONmentalist is "Suppose I agree with your analysis. What can *I* do?" We are all at different levels of awareness and sensitivity to the environment, but whatever stage we're at, we must educate ourselves and get involved. There's a lot that can be done, and here's a very personal list of suggestions to start with.

Inform yourself. There are many helpful organizations, magazines, and books on a wide range of environmental issues. For starters, I'd recommend two old but classic books, Rachel Carson's *Silent Spring* and E.F. Schumacher's *Small Is Beautiful*.

Convince yourself of the reality of environmental degradation. Talk to old people about what the forests, fish, air, birds, mammals were like when they were young. Our elder citizens are a living record of how much the world has changed.

Reexamine some of your most deeply held beliefs. Is steady, continued growth necessary for our well-being? What is the "quality of life"? What is our relationship as a species with the rest of the natural world? What is the ultimate purpose of our government and society? What is progress? Where do spiritual values fit in our lives? The way we answer these questions will determine how we address and act on environmental issues.

Be a conservationist in your daily behavior. Find out where there are depots for glass, paper, chemicals, and metals and start recycling. Use cloth diapers. Store chemical leftovers from the garage, kitchen cupboard, or medicine cabinet for proper disposal instead of pouring them into the sink. Compost your kitchen leftovers. You will be amazed at the reduction in garbage.

Use your power as a consumer. Exert pressure by what you do or do not buy. Praise environmentally responsible companies and criticize ones that aren't. Deep Woods, for example, has an excellent pump spray for insect repellent; why should there be any aerosol sprays when a mechanical pump does the job? Urge stores to phase out use of plastics, ask supermarkets to replace foam containers and other packaging, demand that fast food outlets abandon the incredibly wasteful packaging that adds nothing to the quality of the food. At hamburger chains ask whether they use North American beef.

There is no end to profit-driven waste and pollution. Don't buy leaded gas. Point out the energy waste of open upright refrigerators in food stores. Recycle all glassware and metal containers. Don't buy toothpastes simply because they have color, stripes, or a delivery gizmo that add nothing to deter decay. Don't buy breakfast cereals with little nutritional content just because of a packaging gimmick. Buy organically grown food. This list is endless. Share your thoughts and ideas with others.

Exert your influence as a citizen and voter. Urge all municipal governments to start recycling and set a goal of 60 to 70 percent of all waste. No city or town should be allowed to release raw sewage into rivers, lakes, or oceans. Boats should not be allowed to dump sewage into salt water or fresh water. We have to begin major initiatives to recycle human sewage onto agricultural fields. Agricultural land should not be used for development or landfill.

Press for more legislation with teeth to stop industrial pollution, tighten regulations, and reduce waste. Industry's crocodile tears over excessive costs of pollution control and threatened shutdowns should no longer be allowed to delay implementation of antipollution measures. We need to impose massive fines and prison sentences for corporate executives as well as individuals who pollute. There should be a special police group like the Los Angeles Strike Force to track down and apprehend polluters. Vehicular exhaust should be rigidly controlled. We need massive crash research and development programs on alternative energy, pollution detection and control, and environmental rehabilitation. Governments have to approach environmental matters holistically rather than partitioning them into ministerial departments. Politicians act when they feel the heat of public pressure: letters, phone calls, and telegrams do have an impact. Praise politicians with good environmental records.

Take an active part in elections. Attend all-candidates meetings and ensure that those running have thought about questions of nuclear energy versus alternative energy, atmospheric degradation, pollution, pesticides, and the relationship between profit and growth and environmental degradation. It should be as essential for any candidate to have a serious environmental platform as to be able to read or add.

Support environmental groups. There are many effective environmental groups fighting at different levels. Contact the Canadian Environmental Network through the federal Environment ministry for the list of organizations across Canada and then choose according to your priorities. They need money, support, and volunteer help. My personal list includes Canadian Arctic Resources Committee (CARC), Acid Rain Coalition, Probe International, Energy Probe, Pollution Probe, Canadian Environmental Defence Fund, Greenpeace, and the Sea Shepherd Society.

Think about our children. For the first time in history, we know that our children will inherit a world that is radically impoverished in biodiversity and afflicted with major problems of degraded air, water, and soil. Children should learn that pollution and waste are obscene and represent assaults against them. Surely the opportunity for youngsters to anticipate a rich and full life in balance with the complex community of life on this planet is the reason for society and governments. Profit is not the reason.

This list is far from complete. None of the individual acts will save the world from the impending catastrophe, but involvement changes us, provides us with new insights, leads us into different strategies. It's the *process* that matters. It's the struggle that gives hope.

Teaching the Wrong Lessons

CHILDREN LEARN BY EXAMPLE. THEY WATCH PARENTS AND QUICKLY pick up attitudes from their actions. In spite of the vast expanse of wilderness in this country, most Canadian children grow up in urban settings. In other words, they live in a world conceived, shaped, and dominated by people. Even the farms located around cities and towns are carefully groomed and landscaped for human convenience. There's nothing wrong with that, of course, but in such an environment, it's very easy to lose any sense of connection with nature.

In city apartments and dwellings, the presence of cockroaches, fleas, ants, mosquitoes, or houseflies is guaranteed to result in the spraying of insecticides. Mice and rats are poisoned or trapped, and the gardener wages a never-ending struggle with ragweed, dandelions, slugs, and root rot. We have a modern arsenal of chemical weapons to fight off invaders, and we use them lavishly.

We worry when kids roll in the mud or wade through a puddle because they'll get "dirty." Children learn attitudes and values quickly, and the lesson in cities is clear—nature is an enemy; it's dirty, dangerous, or disgusting. So youngsters begin to wall themselves from nature and to try to control it. I am astonished at the number of adults who loathe or are terrified by snakes, spiders, butterflies, worms, birds—the list seems endless.

Yet for 99 percent of our species' existence on the planet, we were respectful of and dependent on nature. When plants and animals were plentiful, we flourished. When famine and drought struck, our numbers fell accordingly. We remain every bit as dependent today; we need plants to fix photons of energy into sugar molecules and to cleanse the air and replenish the oxygen. It is folly to forget our dependence on an intact ecosystem, but we do so whenever we teach our offspring to fear or detest the natural world. The message urban kids get runs completely counter to what they are born with, a natural interest in other life-forms. Just watch a child in a first encounter with a flower or an ant—there is instant interest and fascination. We condition them out of it.

I see it when my ten-year-old daughter brings home new friends and they recoil in fear or disgust as she tries to show them her favorite pets—three beautiful salamanders. And when my six-year-old comes wandering in with her treasures—millipedes, spiders, slugs, and sowbugs that she catches under rocks lining the front lawn—children and adults alike usually respond by saying, "Yuck."

I can't overemphasize the tragedy of that attitude. For inherent in this view is the assumption that human beings are special and different and that we exist outside nature. Yet it is this belief that is creating many of our environmental problems today.

As long as we have cities and technology, does it matter whether we sense our place in nature? Yes, for many reasons, not the least of which is that virtually all scientists were fascinated with nature as children and retained that curiosity throughout their lives. But a far more important reason is that if we retain a spiritual sense of connection with all other life-forms, it can't help but profoundly affect the way we act. The yodel of a loon at sunset, the vast flocks of migrating waterfowl in the fall, the indomitable salmon returning thousands of kilometers—these phenomena of nature have inspired us to create music, poetry, and art. And when we struggle to retain a handful of California condors or whooping cranes, it's clearly not from a fear of ecological collapse, it's because there is something obscene and frightening about the disappearance of another species at our hands.

If children grow up understanding that we are animals, they will look at other species with a sense of fellowship and community. If they understand their ecological place, the biosphere, then when children see the great old-growth forests of British Columbia or the Amazon being clear-cut, they will feel physical pain because they will understand that those trees are an extension of themselves.

When children know their place in the ecosystem and see factories spewing poison into the air, water, and soil, they will feel ill because someone has violated their home. This is not mystical mumbo-jumbo. We have poisoned the life-support systems that sustain all organisms because we have lost a sense of ecological place. Those of us who are parents have to realize the unspoken, negative lessons we are conveying to our children. Otherwise, they will continue to desecrate this planet as we have.

It's not easy to avoid giving hidden lessons. I have struggled to cover my dismay and queasiness when Severn and Sarika appear holding a large spider or when we've emerged from a ditch covered with leeches or when they have been stung accidentally by yellow jackets feeding on our leftovers. But that's nature. I believe efforts to teach children to love and respect other life-forms are priceless.

Losing Interest in Science

THE MOST IMPORTANT ISSUES THAT THE NEXT GENERATION WILL have to contend with will result from the application of science and technology. An interest in these fields needs to be encouraged. Unfortunately, for many youngsters today, the way science is taught in school turns them off, and too many have stopped taking any science courses midway through high school. It doesn't have to be that way.

When my daughter Severn was five, she asked me to explain how a plant's root had the strength to grow through the ground. Looking out the window, I remarked that the long bump in the road was caused by the tree's roots tunneling under the pavement and pushing it up. Well, this amazed her, and in that instant I reexperienced a moment of discovery that each of my children has shared with me.

Every one of my children has grown up with the weekend nature hike, first in the large woods adjacent to our home and then along the seashore. All my youngsters and their countless friends loved these outings—I have yet to meet a child who doesn't.

They very quickly learn where to hunt for salamanders, the edible mushrooms, robins' eggs, or wood-boring insects. They know from the frogs' songs when to go and look for eggs so that they can follow their metamorphosis. They never have to be told that what they're learning is entomology, developmental biology, or mycology—they just love it for what they get out of it.

And they ask questions that, as a biologist, I find fascinating, questions like: "Daddy, why are your eyes so small?" "Why is Sarika's skin darker than mine?" "Why does Lisa's daddy talk funny?" (he has a cleft lip and palate) or "Can a mommy dog ever give birth to a baby kitten?" These are the kinds of questions that have intrigued many biologists for their entire professional careers and that got many of us into science in the first place.

What happens to that innocent joy in learning about the world around us? As educators design new curricula, I would plead with them not to perpetuate any longer one of the great myths and turnoffs—*the scientific experiment.*

When I was a high school student, we went into a lab and were told what the experiment would be, received a set of instructions, and then were expected to use the equipment to obtain data. Because the emphasis was on the mechanics of doing experiments, we frequently lost sight of the reason for doing them. Without an appreciation of the body of knowledge, insights, and theories that make an experiment definitive, a student can go through a lab exercise like a cook following a recipe.

In high school, the part of a lab exercise most prized by teachers seemed to be the "write-up." We were drilled in the proper steps: define the purpose, describe the materials and experimental methodology, document the results, discuss the implications, and finally draw conclusions. Not only did we have to conform to this protocol, but our reports were often graded on whether we obtained the "right answer."

Having been a scientist now for more than twenty-five years, I can tell you that this is *not* how science is done, and we lose a great deal by teaching it this way. To begin with, there is no such thing as a right or wrong answer—if we knew the right answer beforehand, we wouldn't bother doing research. But even when we repeat a test that has been well documented, the data we get are not "wrong" if they fail to conform to expectations. We may have duplicated the experiment poorly, but the data are all one has.

Those school exercises gave the impression that scientists attempt to solve a problem by proceeding in a linear fashion from A to B to C to a solution. In my opinion, it works much more like my experience in genetics.

In the late 1960s, my lab had demonstrated that it is possible to recover mutations in fruit flies that are temperature-sensitive. Thus, at one temperature such mutants are indistinguishable from normal flies, whereas at just a few degrees warmer or cooler, they die or exhibit a wide array of defects. This sensitivity is a useful property; by controlling temperature, we can determine when and where the defective gene is acting during the fly's development from an egg to adult.

In bacteria and viruses, such mutations represent a single molecular change in the gene product that makes it less stable at certain temperatures. How could we find out whether this was the basis for temperature sensitivity in fruit flies?

We decided to look for a protein defect and focused on muscle, which flies have lots of. What effect would a temperature-sensitive defect in a muscle protein have? We speculated that it might be flexible and permit movement at one temperature but "seize up" and cause paralysis at another. So we chose to look for flies with the trait of paralysis. Then the questions came flooding in. Do we look for nonmovement in larvae or adults? How do we pick out the paralytics from mobile flies? How would we induce mutations? How would we impose the temperature regime? How many flies would we have to screen? Each question had to be answered by trying small experiments to guide us in the proper direction. We had to build special equipment, design genetic combinations, and expand our cultures for a huge test run. All this took months and we were driven to do it by the amused skepticism of our colleagues. Once we got going, we recovered many flies that were dead or sterile or failed to transmit the paralytic condition to their offspring.

After screening more than a quarter million flies carrying chromosomes exposed to powerful chemical mutagens, we recovered a temperature-sensitive paralytic mutation. At 22° C (72° F), the flies could fly, walk, mate—they seemed normal—but shifted to 29° C (84° F), they instantly fell down, completely paralyzed. When they were placed in a container kept at 22° C, they resumed flying before they hit the bottom! It was a spectacular mutant, and when we found it we screamed and danced and celebrated.

But what was its molecular basis? We soon discovered that our mutation (and many others that were subsequently recovered) was a defect in *nerves* rather than muscle.

And how did we write up our results? We riffled through all our records, selected the ones that said what we wanted, and then wrote the experiment up in the proper way: purpose, methods and materials, results, and so forth. The paper was written in a way that suggested that we began with a question and proceeded to find the answer. That's because the report was a way of "making sense" of our discovery, of putting our results into a context and communicating it so that colleagues could understand and repeat (if necessary) what we did. But it conveyed nothing of the excitement, hard work, frustration, disappointment, and exhilaration of the search—or the original reason we started the search for paralytic mutants.

By emphasizing a proper way to do an experiment and to write it up, we create a myth about how research is done. And we lose all the passion that makes the scientific enterprise so worthwhile. I hope the new science curricula don't make this mistake.

Right now, science is being taught on a totally unrealistic model, and unfortunately, for the majority of our students, it's a turnoff very early. Indeed, the word *science* is pretty much of a pejorative by the time they reach high school—it's a subject for the "math brains." And certainly for most teenagers, science is an activity so esoteric that it really isn't relevant to their daily lives. What a sad state, when sex, drugs, and jobs are important to them, and science has a lot to contribute to an understanding of these issues.

The vast majority of teachers who teach the pitiful amount of science in primary schools are very poorly grounded in science, having had perhaps a few hours of lectures in the education faculty a decade or more previously. It's certainly not the teachers' fault, but these days, when we hear so much about the information explosion and the need to get in on the action in high technology, it's tragic that so many children are uninterested in science by the time they reach more highly qualified teachers in high school.

That natural capacity to be excited when discovering things in the world around us is so precious and so easily extinguished that I think political posturing about getting Canada into world-class science is a waste of time unless we devise ways to keep our most talented youngsters interested.

The students I teach in the university are all enrolled for science degrees. They are committed, with specific jobs in mind—medicine, forestry, agriculture—and sadly, most of them have already lost that sense of wonder and joy in learning. They have managed to survive the science curricula but have been pounded into submissive memorizers and grade grubbers. When I returned to the university in 1985, I stopped teaching in classes.

I had always prided myself as a teacher. Teaching to me was a very intimate thing: it meant sharing a part of my life, my thoughts with the class. In the early eighties I felt brutalized by the students' preoccupation with what would be covered on the exam, what percentage of their final grade this material would make, and so on. I feel too old to put up with that anymore.

But I have two youngsters who, like their old man, are crazy about their salamanders, their beetle collection, and their pressed flowers. I don't want to see them lose that. Something must be done about science in the schools.

A Buddhist Way to Teach Kids Ecology

TOKYO TODAY IS A NIGHTMARE OF CONCRETE BUILDINGS AND ROADS crammed with people and motor vehicles. The air is visible and catches in the throat. Rivers are tamed to flow within concrete walls beside the roads. This is human habitat, hostile territory for other species. Urban dwellers here are cut off from the kind of nature and wilderness most Canadians take for granted. But without intimacy with nature, we can confuse crimes against the Earth with economic and technological progress.

Eight years ago, I met a remarkable teacher of Grade 3 students in Tokyo. Toshiko Toriyama told me that most of her pupils are so disconnected from nature that they think that fish live their entire lives in styrofoam and plastic wrap! Sadly, I have met Canadian children who are unaware that hamburger meat or wieners are the muscles of an animal or that the constituents of potato chips and bread once grew in the ground.

I met Toriyama again here in Tokyo. After teaching for thirty years, she quit in order to create a new kind of school based on the writings of Kenji Miyazawa, one of Japan's most famous poets from early in the twentieth century. He left a legacy of wonder at the mysteries and interconnectedness of nature. In Miyazawa's Buddhist universe, everything, from each grain of sand to every raindrop, insect, and plant, is interconnected by fine "threads." The destruction of any part of that elaborate web tears holes in its integrity. This image is the basis of Toriyama's lessons.

Her most useful teaching tool is each child's imagination. In Imagination Class, the children close their eyes and listen to Toriyama's voice, to music, to noises, and to silence. She lets them imagine they are eggs in a praying mantis cluster. They struggle out of the egg case, look out at a new world, and then discover the hazards of predators, the need for food, the search for a mate, and finally, death. Through the life cycle of the insect, children come to experience challenges that are common to all living things. When the exercise is completed, Toriyama told me, the children become very energetic and read and write about their new insights.

"Children take things for granted and don't think about them enough," she said. "For example, they often lose their pencils. So I get them to imagine being a pencil." She takes each component of the pencil—wood, rubber, paint, metal, lead—and gets the children to go back to their origins. With wood, for example, they imagine the forest where trees grow, vicariously live as a tree, and then think how they are logged, transported, and processed.

The children study history through the man who first brought pencils to Japan from Paris a hundred years ago. Eventually, the lessons are reinforced by a visit to a pencil factory. Once they are aware of the true value of their pencils, the youngsters appreciate and take better care of them.

Toriyama teaches the ecology of water by getting the children to be the water of the oceans, evaporate into the sky, rain into rivers and lakes, and become part of their own bodies. One of the most dramatic lessons is about food. She brings a live chicken to class and then kills it, often to the consternation of many children. It is then plucked, cleaned and cut up, cooked and eaten. The children also imagine being pigs, starting as embryos in a womb, then being born, growing, and dying. They go to a farm to see how pigs are raised for mass markets and visit a slaughterhouse to witness the pigs being killed and butchered. It is a powerful experience that reveals our nutritional dependence on other living organisms.

Toriyama is fifty-two and feels desperate about the dire state of the Earth. "At my farewell class, I asked the children how many thought the air would be cleaner eight years from now. No one raised their hand. Then I asked, 'Will it be dirtier?' and they all raised their hands. It was the same when I asked about forests, oceans, and rivers. The children expressed a deep sense of despair."

Referring to environmental degradation, Toriyama said, "We human beings have never gone through this, so our children are having to go through this, without being taught what to do. How painful it is. I want to tell adults to wake up. I want to tell children, 'Don't give up. We have to work together.' Adults work hard to send children to good schools, but what are they struggling for? Is there anything more important than working for the health of the Earth?"

An education that prepares students for a high-tech future and competition in a global economy misses the fact that we are completely dependent for survival and the quality of our lives on the integrity of the planetary biosphere that we seem intent on destroying. Toriyama's students know what really counts.

The System and the Ecosystem

IF THE FUTURE IS UNKNOWABLE, THE BEST STRATEGY FOR OUR young people is flexibility and a broadly based education. Most of all, they should have an intimate connection with nature.

Any child who knows that a dragonfly can fly off before it can be grabbed, observes that a seed will germinate and send roots *down* and leaves *up,* watches a squirrel leap deftly from branch to branch, has envied birds soaring effortlessly, and has followed a caterpillar through metamorphosis into a butterfly understands that no human technology can ever come close to matching living creatures. Our current technological achievements are impressive by human standards, but measured against the scale of life's complexity, they must be seen as crude and superficial. This perspective has to generate some humility.

A child with a love of nature also recognizes that we *share* this planet and that we derive our sustenance from other life-forms. If we forget that packaged eggs or hamburger came from animals, a cotton shirt from a plant, a wooden chair from a tree, then we have lost that connection with nature. In November 1985, I watched a performance that brought that reality back into focus.

I was at a meeting in London on the subject of toxic wastes. After my rather ponderous talk, Jack Vallentyne, an ecologist from the Centre for Inland Waters in Burlington, Ontario, got up. Jack wanders the planet carrying a large globe of the Earth resting atop a backpack. It looks ... well ... unusual. He performs for children around the world by assuming the character of Johnny Biosphere. Like Johnny Appleseed, Vallentyne plants the seeds of ideas in the fertile soil of children's minds. He put on quite a show, transforming an audience of some 700 self-conscious and skeptical adults into wide-eyed children who shouted answers back to his questions.

The most effective part of Jack's show was the way he demonstrated how much we are a part of the ecosystem. He asked all of us to hold our breath for a few seconds and then informed us that we had held gas molecules in our lungs that had been in the lungs of everyone else in the room. Initially, it made us want to stop breathing, but it brought home with a jolt the reality that we share the air with everyone else. Then Jack told us that we all have molecules in our bodies that had once been a part of every single human being who had ever lived in the past three thousand years! (And he didn't even mention animals and plants.) From Jesus Christ to Marie Curie to Britney Spears, we are all linked by shared molecules in the air, water, and soil.

Jack then went on to tell us a story about a Native canoeing on Lake Superior five hundred years ago. It was a hot summer day and he was sweating, so he decided to go for a swim. And as that Native swam in the lake, the sweat washed from his body and was diluted in the lake. Sodium and chloride ions in the sweat diffused through the lake and today, five hundred years later, when we take a drink of water in Toronto, we drink sodium and chloride from that aboriginal man's body long ago.

This is a modern description of a spiritual vision of our relationship with all life on Earth. It is fitting that Vallentyne's target is children. The difference between his approach and mine is worth noting. I tend to emphasize the destructiveness of a value system that sees humans outside the ecosystem and all of nature as a potential resource. I believe we must recognize that the limited vision of science and technology does not give us control, so then we may try to change directions. My operating faith is in the power of reason to overcome cultural values that are generations old. But it's a bleak picture. Vallentyne uses a radically different approach—his message is just as dark, but it's delivered to children who have not yet accepted all our cultural values. He can revel in the unity of all life in the biosphere in a spiritual way that is both uplifting and wonderful. And I think he's on to something.

Why Sterile School Yards Are a Waste

WHEN YOU WERE A CHILD, WAS THERE A SPECIAL PLACE THAT EVOKED wonder, mystery, security? This question was posed by Gary Pennington, an education professor at the University of British Columbia, to two hundred people attending a conference in Vancouver entitled "Learning Grounds—School Naturalization." After an interval of reflection, someone piped up with "a cherry tree," starting a torrent of answers: "a ditch," "a swamp," "my grandpa's garden," "a sand dune," and on it went. Many of those magical spots no longer exist.

It's not cities, shopping plazas, or buildings that make Canada an extraordinary land, but the haunting beauty of the Arctic, the endless horizon of grass and sky of the Prairies, rugged mountains and ancient rain forests, and spectacular autumn leafscapes. These are what inspire our art, poetry, dance, and music and evoke the envy of people abroad, who think of Canadians as people of the outdoors.

But now that most of us live in cities, we have been distanced from nature and must make deliberate efforts to experience it. The Vancouver meeting was an attempt to find ways to bring nature into the lives of urban children by focusing on schools, where they spend a large part of their young lives.

When our children were in primary school, Tara and I would organize field trips to the beach when the tide was very low. It was the first time for many of the children to wander the beach, roll over rocks, or dip into tidepools, even though they live in a city whose beachfront is one of its proudest features. Often the children were afraid to put their hands in the water or touch an anemone or crab. But curiosity invariably overcomes reticence, and they were soon immersed in tidepools, reveling in their discoveries.

That's why school yards are important. They can provide an opportunity to watch the seasons change, observe the succession of plants through the year, and witness the interdependence of insects and plants and birds and soil. Students can see interconnections between air, water, and soil and note the remarkable metamorphosis frogs and insects undergo. In school yards, children can actually grow vegetables and flowers and compost their lunch scraps to learn about the relationship between food and soil.

But school yards are seldom designed for the joy of play and discovery. Instead, they are planned out of fear—fear of litigation, fear of accidents, fear of lurking strangers. These concerns must be addressed, of course, but they should not be the primary determinant of how yards are conceived.

At the meeting, a parks board member told how clover in a school field had been removed because a teacher had complained that children might be stung by a honeybee flying into the classroom. But since honeybees die after stinging, they don't pursue targets, and children can quickly learn to respect them and avoid interfering with their important work. My children's school yard was covered over with coarse gravel that caused far more cuts and scrapes than there were stings when there was grass. The soil is now a toxic wasteland, saturated with chemicals that will retard plant growth for years.

An outdoor environment should be a place of delight and joy, a place of surprise and constant stimulation. What could be more enchanting than a pond filled with tadpoles and chirping frogs, trees with low branches to climb, flowering shrubs and edible fruit trees, and fields of wild plants to attract butterflies and beetles? It's not necessary to sterilize school grounds to avoid allergies, stings, water accidents, and assaults by drunks or perverts.

School land is constantly being taken over for more parking, new portables, sheds for tools, and specialized playing fields. The remaining space should be one of those places children will remember later in life. Right now they are learning that nature is frightening, so we nuke weeds and insects with chemicals; that soil is "dirty," so we cover it with asphalt or gravel; that wild things are tough and dangerous, so we prefer weak, dependent grass; and that children have little need for or few lessons to learn from nature.

Greening of school yards reflects a change in attitude that is vital in a new relationship with the rest of life on Earth. The Evergreen Foundation is one of the organizations attempting to green urban areas.

The Invisible Civilization

IN ORDER TO LEARN FROM NATIVE PEOPLE ABOUT OUR RELATIONSHIP with the natural world, we have to recognize that they exist. To most of us, they are invisible.

A radio reporter is interviewing one of the early settlers in the northern part of British Columbia. The old-timer recalls that when he arrived in the area, "there was no one else around," adding as an aside, "just a *few Indians.*" Now I'm sure the man meant to say that there were no other people of European background in the area, but his remark recalls Ralph Ellison's classic novel *Invisible Man.* Ellison's book was a searing commentary on the consequences of being black in the U.S., the most dehumanizing aspect of which was being invisible to the white majority.

On May 5, 1988, B.C. minister of forests Dave Parker met with a First Nations group from the interior of B.C. to continue negotiations on the future of the Stein Valley. Chief Ruby Dunstan of the Lytton Indian Band listed a number of grievances in which their requests had been repeatedly ignored or slighted. She ended with the plea, "Stop treating us as if we are *invisible!* We're human beings too."

Parker's reaction was astonishing. He took umbrage and huffed, "I deeply resent what you've just said. There's reverse racism too, you know." Chief Dunstan's plea to be treated with dignity was grotesquely twisted into a ministerial insult.

As in Ellison's novel, Native people in Canada don't seem to exist except as government statistics. When Europeans arrived in North America, the continent was already fully occupied by aboriginal peoples with rich and diverse cultures. Yet Canadians refer to the French and English as the "two *founding* races." This denial of the very existence of already thriving Native people is reinforced by their cursory description in our history books. Our governments have systematically oppressed and exploited First Nations people, destroyed their culture, and denied or opposed their right to claim aboriginal title.

The history of Native people after contact with Europeans is a tragedy that continues to the present. Hunting-and-gathering nomads were forced to give up their traditional way of life for permanent settlements. Children were exiled from their remote homes to be educated in urban centers hundreds or even thousands of kilometers away. There they were taught to be ashamed of their culture and were punished for speaking in their own language. And we outlawed some of their most important, even sacred, cultural activities—the potlatch, the drums, dancing, religion, Native medicines, even hunting and gathering.

Their original population ravaged by disease, dispossessed, forced to leave the bush and to abandon ancient traditions, aboriginal people today occupy the lowest rung in social standing in Canada. It is hardly surprising that alcoholism, unemployment, suicide, sexual abuse, and crime afflict many Native communities across Canada. And these negative images are only reinforced by their constant repetition in reports by the electronic and print media.

I have recently watched three films about Canadian First Nations that had been commissioned by the aboriginal people themselves. What a difference they are from the usual stereotypes in the media. The films allow Native people to talk about themselves and their own ideas, obstacles, and goals. The reaction of media people who screened the films for possible television broadcast is instructive. "Beautiful footage, but too one-sided" was a standard response. When a Native person in one film talked about the urgency of conserving salmon because the fish are at the core of his culture, a television executive snorted, "Are they trying to tell us they know something about ecology?" Another remarked, "If you leave it to the Indians, there won't be any salmon left." Another person suggested, "The women are too fat and the men look too white." When the narrator said, "Along these banks are remains of ancient civilizations dating back nine thousand years," one man interjected, "I don't think you should call them civilized, because they were nomadic."

Underlying these responses is the assumption that people in the media know what reality is and how Natives should be portrayed. But there is no such a thing as objective or balanced reporting in the media. Everyone in the world is molded by heredity, personal experience, and cultural milieu, factors that shape not only our values and beliefs but the very way we perceive the world around us. It's easy to confirm this—just talk to an Iranian and an Iraqi, a Northern Irish Catholic and a Protestant, a South African white and a black, or an Israeli and a Palestinian about events that both in each pair have witnessed.

The point is that we edit our experiences through the lenses of our personal worldviews. That creation of subjective reality doesn't suddenly stop when people become part of the media. Human beings cannot help but impose their personal priorities, perceptions, and biases on their reporting because *that is all they know.* Implicit in our newscasts are all kinds of biases—we just don't recognize them as such because they happen to be the dominant view in society. But ask a Soviet or Chinese visitor and they'll see the biases immediately.

As long as Native people are not a part of the media, they will be portrayed as the rest of society chooses to perceive them. At the very least, the media ought to be willing to let members of the First Nations present reports from their point of view, if only to make them a little less invisible.

JOURNALISTS OFTEN DO HOLD UP OBJECTIVITY AS A HIGH GOAL IN reporting. A few years ago, one of the *Globe and Mail*'s most respected reporters chastised me for publicly supporting a political party, because, in his opinion, it compromised my objectivity as a broadcaster. When I signed a letter supporting an antinuclear petition, I was informed that some in the Canadian Broadcasting Corporation management felt I had lost credibility on nuclear issues.

If there really is such a thing as "objective journalism," then there would be no difference in reporting by women or members of visible ethnic communities. Nor would it matter that there is a preponderance of upper-middle-class white males doing the reporting, as long as they were "objective."

If most reporting is truly objective, then there is no legitimate claim for the necessity of Canadian print or electronic media. If reporters simply observe and transmit objective reality, then the source of news ought not to matter. Of course this is ludicrous. The reason we value the CBC, the National Film Board, and Canadian magazines and newspapers is that they present perspectives from within this country's culture. None of us can escape the limitations of our heredity and personal and cultural experiences. There's no such thing as objectivity.

There is plenty of evidence to show that complete objectivity does not exist even in that most rational and objective of all activities, science. Harvard's great science popularizer, Stephen Jay Gould, wrote a marvelous book entitled *The Mismeasure of Man*. In it, he documents the history of the scientific study of the human brain and shows how existing beliefs and attitudes affect not only the kinds of questions asked and experiments conducted but also the way the results are interpreted.

So, for example, when it was believed that brain size was correlated with intelligence, scientists obtained evidence that the brains of blacks were smaller than the brains of whites. Decades later, when Dr. Gould measured the cranial capacity of those very same skulls, he found there were no statistically significant differences. By then it was also clear that brain size alone is not an indicator of intelligence.

Similarly, Dr. Gould describes how scientists once believed that intelligence could be pinpointed to a certain part of the brain. And, sure enough, when comparisons were made, that part of the brain in women was found to be significantly smaller than it was in men. Years later, when it was known that that particular part of the brain had nothing to do with intelligence, a reexamination of the data revealed that the differences were not significant.

We acquire genetic and cultural "filters" through which we perceive the world around us. I was struck with the power of those filters in 1987 when I visited the Stein River Valley. I was flown by helicopter into the valley with my host, a Lillooet Indian Band member. He pointed out the burial grounds of his ancestors, the battle site between his people and the neighboring tribes, and their ancient hunting grounds. Our pilot told me that a week before he had taken a load of foresters over the same area, and all they spoke of were the number of jobs, the years' worth of logging, and the enormous profits those trees represented. The foresters and the Native people were looking at the very same place, but what they "saw" was worlds apart.

Now you see why Native people must have a means of seeing themselves through the lenses of their own values and culture; otherwise, they will live only with the fabrications of non-Native journalists.

But I have a far more selfish reason for supporting them. I believe that as North Americans explore the spiritual values of Eastern religions and African cultures, we ignore an important perspective right here in our midst. Canada's First Nations have a view of their place in nature that is very different from the non-Native view, and in spite of the way they have been brutalized over the centuries, they have hung on to those differences. Through a Native perspective, non-Natives can measure and reexamine our own assumptions and beliefs. It is only in having a contrasting view that we can truly recognize our strengths and deficiencies. To do that, we have to abandon the myth that there is some high form of objective reporting and acknowledge our inherent and inescapable biases.

Haida Gwaii and My Home

IN OVER A DECADE OF ACTIVITY WITH ABORIGINAL PEOPLE IN North and South America, Asia, and Australia, I have had opportunities to think back on my own society from a very different perspective. In 1990 my family and I spent the last week of summer on the northern tip of a land Haida people have occupied since beyond memory. They call it Haida Gwaii. We newcomers to this place have named it the Queen Charlotte Islands. Perched on the western rim of Canada within spitting distance of the Alaskan panhandle, this remote archipelago is such a rich storehouse of animal and plant life that some call it Canada's Galápagos Islands.

We stayed in a new longhouse at the site of an ancient village called Kiusta, just east of Yaku, where old poles still stand as silent reminders of the great civilization that once flourished here. Across the neck of a peninsula, three new Haida longhouses at Taa'lung'slung guard the pristine beaches. Here the most dedicated workaholic has to adjust his or her pace to the rhythm of the surroundings. Here one has time to reflect.

My spine tingled to walk on land that once echoed with drums and songs of a vibrant people but where now only twisted trees and dim outlines in moss reveal the poles and longhouses. Not long ago, the Haida occupied dozens of settlements and hundreds of temporary sites throughout the islands. How could they have survived the catastrophe caused by the smallpox epidemic that wiped out 80 to 90 percent of the people in a matter of years in the late 1800s? The strongest were just as susceptible to smallpox as children and elders. How did the few survivors keep from going mad as the fabric of their communities was so cruelly destroyed? Today, smallpox has been declared extinct, and miraculously, the descendants of the Haida survivors still occupy two villages (Masset and Skidegate) and are reasserting their culture and presence. The new longhouses attest to that.

Across the narrow strait separating Kiusta from Langara Island, three huge floating lodges for sport fishers are anchored. To manage the salmon resources of the islands, the Haida have formulated a plan that calls for a reduction in catches by the sport fishery. They have been supported by most of the small lodges and commercial trollers. So in the matter of fish, there is an opportunity for all islanders to work out a plan to sustain the yield of a renewable resource.

From Taa'lung'slung, the Pacific Ocean stretches all the way to Japan. Yet even here, the beaches are littered with the familiar detritus of modern society—plastic in all forms, glass bottles, floats, rope, and much more. It is a reminder that the planet is a single entity and that we don't get rid of garbage; we only shift it around.

Around Kiusta, the biological abundance in the forests and the water is overwhelming. The trees of the rain forest are massive and untouched. The bays are jammed with humpbacks (pink salmon), waiting for rains to swell the rivers, while the dog (chum) salmon are just arriving. The surf teems with tiny shrimplike creatures that are at the base of the zoological food chain. Out at sea, huge clouds of birds—puffins, auklets, gulls—gorge on immense schools of needlefish, while black bass and other fish flop out of the water in pursuit of the same prey. And, of course, the prized chinook salmon are being caught in large numbers by sport fishers from the lodges.

In spite of the sense of limitless abundance in Haida Gwaii, changes are noticeable. In the middle of the island chain, large clear-cut areas in the forest are visible everywhere. Abalone are being rapidly depleted, and millions of giant clams called geoducks are being blasted out of the ocean bottom and shipped to Japan, even though scientists know virtually nothing about their biology and reproductive needs.

Salmon, birds, and many marine mammals feed on herring, yet the small fish is exploited in a most wasteful, shortsighted way. Spawning herring are netted by the millions just for the females' eggs, then the carcasses are rendered for animal feed or simply thrown away and the eggs are sent to Japan. The Haida traditionally harvest the roe (*gow*) after it is laid on kelp so that the adults can return to spawn again. Why, then, is roe herring fishing allowed? It makes no sense at all.

Motoring around Langara Island, we spotted a tall, thin spout and then the immense tail of a sperm whale as it made a deep dive. It reminded me of our July trip to the Khutzeymateen Valley to watch grizzly bears. On the boat out, we had encountered a pod of killer whales. When one of the animals surfaced after a long dive, my ten-year-old daughter burst into tears. "That whale went so far on one breath of air," she wailed, "and we coop them up in such tiny tanks in the aquarium. It's cruel!"

Looking up at the spectacular display of stars in the clear night sky, I thought about the recent visit of relatives from Cleveland. They had stayed at our cottage, and the teenagers were astounded to see the Milky Way for the first time in their lives! They were equally impressed with being able to drink creek water without worrying about its purity. What kind of world and expectations are we leaving for our children?

Haida Gwaii's remoteness is a great advantage to the people who live there. Their connection with and respect for land and sea are essential because they still depend on them. Cooperation and sharing are a vital part of the communities, and the people have a chance to live within the biological productivity of that enchanted place. For city dwellers, a visit to the islands provides a chance to rethink our values and priorities.

Reflections While Backpacking

ON THANKSGIVING WEEKEND IN 1995, MY FAMILY PASSED UP TURKEY and all the trimmings to backpack up the Stein Valley near Lytton, British Columbia. For more than a decade, a rancorous battle has pitted the logging community against people of the First Nations and environmentalists over the fate of this valley, the last intact watershed in southwestern B.C. To date, the logging road has been kept out.

Exactly five years earlier, we had made the same trip, so we were anxious to see whether much had changed. To our joy, nothing had been perceptibly altered in the interim. The river was full of pink salmon, nicknamed "humpies" from the big lump on the back of the hook-nosed males. These animals had fought their way past obstacles of hooks, nets, pollution, and human development. And here they were gathered in pools to rest, battered and spotted with white patches of fungus that would devour their carcasses. Driven by the imperative to reach the spawning beds before their genetically programmed deaths, they still clung to life and possessed the power to leap barriers in the river.

Within a few hundred meters of the trailhead is a rock wall covered with ancient pictographs that reminded us that this was a sacred valley for the aboriginal people whose descendants still live in the territory. The steep stone faces of the valley are pocked with caves, some of which are burial sites. During the annual Stein Festival, held to celebrate this watershed, Native people often referred to the place in reverential terms as a temple, a sacred place, a sanctuary.

Puffing up the Devil's Staircase, I try to ignore the pain of cramped muscles by concentrating on other things. I think of the president and CEO of a major forest company, whose letter to local mayors had described environmentalists as "anti-everything." Environmentalists are often dismissed as too "depressing," "pessimistic," or the "messengers of doom and gloom." They are castigated as "radicals" or "ecoterrorists." Nothing could be further from the truth. What is truly depressing are the people who so lack understanding that all they can do is mindlessly push for what has clearly failed: more economic growth, regardless of the social and ecological costs.

Environmentalists are the true conservatives—they want to save, protect, and slow down the pace of life. People who want to rush into wilderness areas, like the Stein, that have never been studied or inventoried and trash them without regard to whether they can be sustained or restored are the real radicals and ecoterrorists. I reflect on the words of the B.C. biologist Bristol Foster, who described the struggles to protect areas like the Stein: "In saving wilderness, every victory could be temporary, every defeat is permanent."

Environmentalists are not against everything; they are for the most important things on this planet, things that keep us alive and wealthy. They are for clean air, clean water, clean soil, and a diversity of creatures all over the world. Environmentalists work for local communities that are sustainable, for economies in which full employment, security, and spiritual needs are paramount. And as they work to protect wilderness areas, they celebrate the wondrous variety of life-forms that share the planet with us.

On arriving at the start of the trail, we found eight parked vehicles, yet in three days of hiking we hadn't seen a sign of litter or human debris. For those three days, my wheezing from a newly acquired bronchitis disappeared in the clear air. As we drank the pure, frigid water from the river, it was like a sacrament. There was an overwhelming sense that the salmon, the dazzling fall leaves of the alder, the autumn mists sliding over the mountains confirmed the timelessness of nature's rhythms, which are so easily forgotten in urban areas. By being there, we were giving thanks for this special place and other special places on the planet.

LEADERS, ROLE MODELS, AND SUCCESS STORIES

For most of us who struggle against ecological destruction, the picture is bleak. We must deal with population growth, overconsumption, toxic pollution, species extinction, climate change, marine depletion, and more—a distressing litany of degradation. Fear can be a powerful motivator for immediate bursts of response. A burning house, a car crash, or a flood demand that we react instantly for survival. Fear evoked by the threat of an impending ecological disaster can motivate us for a while, but it is difficult to sustain. Hope can carry us on for the long haul.

Two radio series that I was involved in (*It's a Matter of Survival* in 1989 and *From Naked Ape to Superspecies* in 1999) evoked a large response. But each time, the letters, e-mails, and phone calls demanded to know what could be done. So I suggested to Holly Dressel, my coauthor, that we search for examples of attempts to get onto a sustainable path by individuals, companies, organizations, and governments. As we started to research, I said to her: "I hope there are enough examples for us to get a book." To our amazement, within months it was clear we could have filled volumes. The book was called *Good News for a Change: Hope for a Troubled Planet,* and it became a number 1 best-seller in Canada and Australia. The good news is that there is a lot of good news in both rich and poor nations, from individuals to corporations and governments. The bad news is that most politicians and businesspeople seem determined to carry on with business as usual instead of recognizing the challenges and opportunities in seizing the new path that is being opened.

The stories in this section represent a few of the examples I wrote about from time to time over three decades. They hint at the possibilities that we went on to describe in greater detail and variety in *Good News for a Change.*

The New Leaders

IRONICALLY, THE PEOPLE WHO TODAY ARE OFFERING VISION AND leadership come from groups that have traditionally been powerless and disenfranchised: the Third World, women, youth, elders, and indigenous people.

1) The Third World. In poor countries, 4.4 billion people have had to eke out a survival on less than 25 percent of the planet's resources. The uncertainty of their circumstances is a direct cause of their exploding numbers, because for them children are the only form of old age insurance. But this attempt at security has only intensified the misery of people in the Third World, because the developed world leaves them a pittance on which to live. In addition, industrialized countries have exploited poor nations by dumping outmoded or banned technologies like pesticides and toxic wastes on them.

   But today, the fate of all people in the world hinges on the future of the Third World. If, for example, people in developing countries decide to copy our lifestyle of high consumption and waste, we will all be left with crumbs to fight over. If India and China go ahead with plans to industrialize using their extensive reserves of coal, the level of warming gases in the atmosphere will increase dramatically. China's plan to put a refrigerator in every home by the end of the century has huge implications if ozone-depleting CFCs are used as refrigerant instead of the more expensive alternatives. Yet former U.S. president George Bush rejected India's suggestion that the rich countries contribute to a superfund for the Third World to finance such projects.

Similarly, the future of tropical rain forests depends on our willingness to relieve countries of their debt. While still Brazil's environment minister in 1990, José Lutzenberger made a precedent-setting announcement that the country would consider preserving forests if foreign countries paid for it.

Papua New Guinea announced a two-year moratorium on all new logging licenses in July 1990 to try to save the forest. But the U.S. $70 million in forgone logging income had to be made up by rich countries. What are tropical rain forests that hold more than 50 percent of all species worth to us? Will we who have been so profligate with our own forests choose to pay so that the poor countries will not repeat our mistakes? The Third World makes us face up to it.

2) Women. Over half the world's population has been denied access to the competitive, hierarchical, and patriarchal power structures of government and business. Women have a radically different perspective from men, one that is characterized by caring, nurturing, sharing, and cooperation, the very traits that will be needed to stave off an ecological catastrophe. It is not an accident that women are so prominent and disproportionately represented among the leaders and the rank-and-file of environmental groups around the world.
3) Youth. Those with the most at stake in decisions being made now by governments and in boardrooms of business are youth. After all, they are the inheritors of what will be left. Consequently, they cannot afford to wait until they reach the age to vote. The Environmental Youth Alliance (EYA), an umbrella organization of high school environmental clubs, now boasts more than seventeen thousand members. They have been holding conferences in cities across Canada and attracting hundreds of participants. The EYA has become an international group with dozens of potential member clubs in Australia, and there are plans to expand into other countries, including the United States.

The sense of power and the optimism that youth can change the world are contagious, and young people will be a formidable force. Not only are they informing politicians and business leaders that they want change, but they are recruiting their peers and exerting influence on their parents, too.

Youth cannot be ignored.

4) Elders. In our rapidly changing society, human beings become obsolete as we push them aside when they get old. Yet never before have we needed the experience and perspective of our older citizens so much. They have been through the game of life, know what the rules are, and often realize how ephemeral and irrelevant they are. Retired generals and admirals who opposed nuclear war were powerful voices for the peace movement because they had credibility as former participants in the dangerous game of war.

   We need a group like Retired Company CEOs and Presidents for the Environment to provide the leadership needed to change the priorities of business. Elders are invulnerable to the normal constraints and pressures that inhibit many of us. British Columbia's wonderfully outrageous Raging Grannies score many points because they know they have nothing to lose.

5) Indigenous people. In spite of generations of genocide, oppression, and exploitation, pockets of aboriginal people around the world retain a relationship with the land and a connection to other life-forms that is critical if we are to learn to live in balance with the rest of the living world. These remaining islands of indigenous perspectives are priceless because they can show where we have gone wrong and what we have to regain.

It is time for these disempowered groups to coalesce into an irresistible force for change on the planet.

Village Power Wins Victories in India

HURTLING ALONG A ROAD IN WHAT SEEMS LIKE A SUICIDE RUN, I pray there are Hindu gods to look after foreigners. India is like a different planet, where one's every assumption, value, and belief simply has to be suspended. I'm here for a special program on dams for *The Nature of Things.*

Like many other nations in the less-industrialized world, India has been beguiled by the twentieth-century illusion that bigger is better and that what is modern is superior to ancient traditional ways. This attitude has been encouraged by agencies such as the World Bank.

The Narmada is the largest river flowing west in central India. It supports rich forests and wildlife, as well as tens of thousands of tribal people, who continue to live off the surrounding land. But its most potent value to Indians is spiritual: many consider the Narmada even more sacred than the Ganges. The modern perspective views the river in economic terms, however.

With forty major tributaries, the Narmada River basin drains water from an area of almost 100,000 square kilometers (39,000 square miles). For decades, proposals had been made to harness the flow for drinking water, but especially for power and agricultural irrigation. So since the 1960s, the Indian government has pursued a plan to build two superdams and thousands of major and minor projects on the Narmada. The scheme will be the largest irrigation strategy in the world and will affect 12 to 15 million people in four states. There will be enormous economic, social, and ecological costs.

This is a country where the poor and powerless have always been pushed around by those with wealth and power. But since arriving, I have learned of two remarkable people who have given a sense of power to those at the bottom of the economic and political pile. Mehda Patkar is a Hindu woman who learned of the huge proposed Sardar Sarovar Dam, which will flood hundreds of villages of tribal people. Since the mid-1980s, she has single-handedly galvanized the inhabitants of the villages into action by informing them of the government's plans. Walking hundreds of kilometers from village to village, Patkar marshaled opposition to the dam so that by the nineties, unprecedented public demonstrations involving tens of thousands of protesters were held. Thousands have been arrested, and in 1993 police shot and killed a teenage boy. The protests created so much pressure that the World Bank eventually reneged on its promised loans.

In spite of the expense, the Indian government continued with the dam building. Opponents contended that the dams would be ecological and social disasters. But the most potent criticism was that most of the water would be used for large-scale irrigation of cash crops, which would enrich only wealthy landowners and big companies. Yet in the same part of the country, another person provided an alternative strategy that works.

During the 1965 war with Pakistan, Anna Hazare was driving a Jeep in a convoy that came under fire. Everyone except Hazare was killed. He decided that he had been spared to be reborn again and so dedicated his second life to serving the people of his village, Ralegan Siddhi.

Returning home, Hazare found the people and the land in a terrible state. The main sources of revenue were forty distilleries and a tobacco industry. Hazare wanted an alternative to the alcohol and tobacco industries and looked to the village's agrarian roots. But groundwater had been depleted, leaving little for crops. He recognized that when it did rain, the runoff carried away the meager soil. To Hazare, soil is the life of the village and the village is the unit of survival that must be protected at all costs. So he began to work on "watershed development," with the aim of keeping both soil and water in the community. That involved planting trees and crops that would retain the water and digging a series of horizontal pits that would collect water and slow its flow down the hills. Within four years, Hazare could point to dramatic results—decreased soil erosion, recharged groundwater, water flowing in once-dry riverbeds, and increased crop yields.

Today the distilleries are gone. The village is surrounded by lush green fields and actually exports water to neighboring communities. Hazare's work has been recognized, and he is now coordinating the application of his approach to watershed development in three hundred other villages.

All over the world, people like Hazare and Patkar are rallying support for community-based ideas and technology and taking charge of their own destiny. It's an inspiring lesson for us in the so-called developed world.

Update

Under pressure from environmental and social justice groups, the World Bank established a committee to assess the ecological, social, and economic impact of Sardar Sarovar. Headed by Bradford Morse and Tom Berger, the committee released its report, "The Independent Review—Sardar Sarovar Project," on January 15, 1994. The report was a severe indictment of the claimed benefits of the dam and concluded, "the wisest action would be to step back from the project." The World Bank abided by the recommendation and declined to fund the dam. This decision was unprecedented for an organization enthralled with megaprojects and almost messianic in its faith in development. India found funding elsewhere and continues with the dam. Later, India announced it had a vigorous nuclear program and detonated a series of nuclear bombs.

A Heroic Shepherd

THERE AREN'T MANY TRULY WILD PLACES LEFT ON EARTH WHERE human beings live as our species has for 99 percent of our existence. In parts of the Third World, where the last remnants of large tracts of tropical rain forest survive, it is possible to encounter the lush diversity of living things and some of the last survivors of a hunting-and-gathering way of life. All are at risk and most are falling before the voracious demands of the global economy. On the island that was once called Borneo and that is now a part of Malaysia dwell the tiny band of surviving indigenous people known as the Penan. Their plight has recently become widely known only because of a remarkable Swiss shepherd who lived as a Penan for six years.

Short, deceptively frail-looking, with hair shaved close to his skull, and wearing wire-rimmed glasses, Bruno Manser is an unlikely candidate for hero. But to those who know him, he is already a legend. Born in 1955, Bruno Manser grew up in Basel, Switzerland, fascinated with the way people lived in the past. So he became a shepherd and cheesemaker, living a simple life and avidly reading about indigenous people in other parts of the world. Manser learned about the Penan, an isolated nomadic people who live by hunting and gathering in the rain forests of Sarawak. In 1984, he set out for Southeast Asia to contact the Penan so that he could live with them for a while.

Canadian ethnobotanist Wade Davis says he has never met a people with as profound an understanding of the pharmaco-chemistry of plants as the Penan. They know the oldest, most biologically diverse forest on Earth in a way that scientists never will. It is the accumulated observation and insight of people whose lives have depended on that knowledge for tens of thousands of years. Hunting with deadly accurate blowguns, eating sago palms, and living on elevated platforms, the Penan live a simple nomadic existence. The loss of their forest will mean the loss of nomadic ways and of a body of knowledge that connects us to nature and our past.

But nowhere is too remote for the insatiable demands of industry and global economics, and the forests that are home for the Penan are irresistible. Sarawak is the world's largest exporter of unprocessed tropical timber, two-thirds of it to Japan. Sarawak collects over $5 billion annually for the sale of timber, but little reaches those who need it most, and meanwhile, the forests are irreversibly destroyed. Most of Sarawak's indigenous people have been settled in shantytowns, where they become impoverished replicas of "civilized" people. There are only sixty-two nomadic Penan families left.

The fate of the Penan will reveal how much we care for cultural and biological diversity. Is there room left for values that do not require the accumulation of wealth, the destruction of nature, and the dominance of humankind? That is what Bruno Manser wants to know.

In 1985, after living with the Penan for a few months, Manser was taken into custody by the Malaysian government because his visa had expired. Knowing he would be deported or jailed, he escaped by diving off a police boat into a river and swimming away. Thus, he became a fugitive. For nearly six years he lived as an outlaw with a price on his head, eluding a massive search by the Malaysian government, which regarded him as an agitator and an embarrassment. He helped Natives organize blockades of logging roads, but as Manser watched the Penan being threatened and jailed, he realized they needed support from other countries. He began to tell the story to the foreign press.

In May 1990, in an adventure worthy of Indiana Jones, Bruno Manser was disguised and illegally smuggled out of Malaysia. He remained determined to spread the word to save his people. In response to his plea for help, the Sarawak Circle, an international network supporting the Penan, was organized by the Western Canada Wilderness Committee in Vancouver. The circle called for a moratorium on logging in the Penan forest and the establishment of a Biosphere Reserve for the Penan. It recommended that Sarawak wood products be boycotted and that we inform Malaysia that we admire their national park system and hope that the Penan forests will also be saved. Manser believed that Malaysia and countries like Japan that are actively cutting and importing tropical timber from the shrinking primary forests of the world must be made aware of international concern for the future of those forests and their human inhabitants.

The Malaysian government has proposed the establishment of a Biosphere Reserve, in which the forest would be preserved for the Penan and other tribal people. But the Sarawak state government has complete authority over its own "resources," and the bulldozers and saws have never stopped. Logging interests, which are concentrated in a few families and companies, have enormous political clout and reap huge profits. To them, the Penan are a nuisance.

Engaged in speaking all over the world on behalf of the Penan, Bruno Manser continues to work on a book about his experiences and is a consultant with a Hollywood film company that is planning a movie based on his life. It is inspiring to encounter a man like Bruno Manser and to realize there are ecoheroes all over the world.

Update

Bruno Manser eluded the Malaysian government for years as he organized Penan to protest the logging of their lands. After he left Sarawak to marshal international support for the Penan, Manser founded the nonprofit Bruno Manser Fonds to support the cause of indigenous peoples.

It is said that from time to time, he would sneak back into Sarawak to visit his adopted people. On one occasion in 1998, to attract press attention, he flew a motorized hang glider and landed on the lawn of the Chief Minister of Sarawak. He was immediately arrested and deported.

On May 22, 2000, Manser crossed the border from Kalimantan in Indonesian Borneo into Sarawak. A few days later, he sent a letter to his girlfriend from the border town of Baria. He was never heard from again.

In December 2000, seventeen Penan leaders released a letter announcing that after exhaustive searches, Manser could not be found but that his fighting spirit would live on.

On May 23, 2001, a year after Manser vanished, a totem pole was raised in Switzerland to honor his memory. On August 28, 2001, after a fruitless search led by his brother, Bruno Manser was formally acknowledged to be dead.

Grass-Roots Groups

POLITICAL AND INDUSTRIAL LEADERS ISSUE ENDLESS REPORTS AND press releases about their attempts to balance the economic imperatives with environmental responsibility. But since all three major political parties in Canada believe in maintaining endless economic growth that is a direct cause of environmental destruction, none of them has seriously addressed environmental problems. What is needed is a grass-roots environmental movement so powerful that all of society, including politicians, will be transformed. Such a movement is indeed growing.

As we move away from an ecologically imbalanced way of living to a more harmonious relationship with our surroundings, the big changes will be in the minds of individuals and their communities as they redefine priorities, values, and lifestyles. There are signs that it is happening. Let me give you three examples.

In October 1989, Jeff Gibbs, a twenty-two-year-old student at the University of British Columbia, established the Environmental Youth Alliance (EYA) to link high school environmental groups. EYA connects groups through a newspaper featuring stories by students and invited experts. It also sponsors gatherings at environmental conferences and arranges trips to wilderness areas.

Gibbs's own story is an advertisement for EYA. Raised as a city boy, at fifteen he experienced a fundamental shift in perception while on a canoe trip through the Bowron Lakes in central British Columbia. "Until then," he says, "I always thought human beings were at the top of the heap. But out there, I was overwhelmed with the power of nature and how puny I was."

The next year, in 1984, that spiritual revelation took him to the Queen Charlotte Islands, where "I realized that nature is incredibly complex and runs on its own agenda. If humans weren't there, it wouldn't make a bit of difference. I was blown away by the power, the mystery, and the beauty of it all."

When a battle broke out over proposed logging on Meares Island off the west coast of Vancouver Island, Gibbs started an environmental group called the TREE Club (Teenagers Response to Endangered Ecosystems) in his school. About thirty students joined, and the first thing they did was to collect the names of every elected member of the federal and provincial governments. Each student then chose about twenty names and wrote a personal letter to each by hand, citing statistics and asking them to save the forest. Replies, including one from the prime minister, began to pour into the school. The students were able to tally those for and against logging Meares and focused their attention on the undecideds.

Later, the youngsters ordered three thousand buttons saying Save South Moresby, a contentious area in the Queen Charlotte Islands. The buttons cost 20 cents apiece and were sold for $1. Money poured in to supplement what was raised by bake sales and car washes, in all about $7,000. The TREE Club gave some of the money to the Western Canada Wilderness Committee to print five thousand newspapers about South Moresby, and the students then helped to hand-deliver them.

The TREE Club organized a slide show, which they showed to schoolmates, parents, and the general public. And it made more money! For two months, club members knocked on doors to talk to people about the future of South Moresby, covering over five thousand households.

This is power at the grass roots. EYA will link high school groups across Canada and encourage students to get involved by forming their own environmental clubs. After all, it is their world that's at stake. This movement has swept the country and put a lot of pressure on adults. Listen to what students are thinking.

A Grade 7 class in Ajax, Ontario, wrote: "We'd like to know what will become of the world when we are adults. What will happen if nothing is done? How will we stop the carelessness?"

Vernon, British Columbia: "So far we've raised about $70 in a bake sale, and next Monday, a lady from our recycling depot is going to talk to us about sorting recyclables so we can get people volunteering at the depot. Did Nancy tell you about her idea of going to McDonald's and either bringing china plates or asking for them?"

And from Concordia University in Montreal: "I formed, along with some other students, a recycling committee to try to create awareness about the severe garbage problem. I will be attending a meeting with the mayor of Montreal and representatives from the ministers of environment, communication, and transportation to listen to what they have to say so I can return to school and report the meeting to my fellow students."

This is not a passing fad. These young activists are going to become more insistent and vocal in their demands. Interestingly, a large majority of EYA members and participants are *girls.* And women are leading the moves to change the way we live. Take Andrea Miller.

Miller lives in a West Vancouver high-income neighborhood. She calls herself an environmental homemaker, concerned about the problem of garbage. When she learned that Vancouver's garbage was going to be exported north to Cache Creek, she was galvanized into action. She decided there would be no garbage crisis if there was no garbage. And she has been able to reduce her family's garbage output to under a bag a month. It's a heroic achievement that takes far more than just composting all organic waste and recycling cardboard, paper, glass, metals, and so on.

It requires a major shift in personal priorities, attitudes, and behavior. For example, she does not buy anything that has plastic wrapped around it, she carries her own cup everywhere, and she uses traditional cleaners like borax, lemon, and salt instead of detergents.

In January 1989, Miller began knocking on doors in her neighborhood and inviting people to come over to her place with a friend. She enthusiastically showed her guests simple ways to reduce garbage output. With less garbage, she reasoned, there will be less need for huge incinerators or dumps in other communities and all kinds of related benefits of saved energy, reduced pollution, and conserved resources.

To date, Miller has held dozens of coffee klatches in her home and now gives at least three talks a week to groups ranging from nurses to schoolchildren. She has inspired others who have formed a group called WHEN (Worldwide Home Environmentalists' Network), which offers advice and help to people who want to be more environmentally responsible.

The third example of grass-roots change started when neighbors in a midtown section of Toronto began to meet and discuss the radio program *It's a Matter of Survival.* Responding to the urgent message of the series, they invited all people on that block to meet. At that first meeting, fourteen people showed up. Calling themselves "Grass-roots Albany," they quickly agreed to a set of goals: (1) They will work to clean up the environment around their homes and neighborhood. (2) They will enter into a vigorous and continuing correspondence with politicians at all levels of government. They will not be deterred by form letters and will persist with follow-up letters demanding dialogue. They will vote on the basis of environmental leadership of candidates rather than for parties. (3) They will pursue individual environmental goals with the support of the group. Some of the projects embarked on under this category are interesting. A university teacher will try to stop the use of styrofoam containers at his institution. A member will find out how clean the city's water is and research the best filtration system. Another member is going to pressure the Board of Education to begin a massive tree-planting program on all school property to start reforesting the city, while someone else will compile lists of practical tips on how to live in an environmentally responsible way at home.

One of Grass-roots Albany's recent projects is called "Preserving the Urban Forest." If we think of forests as the complex communities of organisms in an untouched watershed, then "city" and "forest" seem a contradiction in terms. But urban trees can also be thought of as a different kind of forest that nevertheless plays an important role in our lives.

A couple of years ago, the great Haida artist Bill Reid suggested to me that environmentalists should hold a rally "to protest the clear-cutting of Kerrisdale" (an upscale part of Vancouver). While said playfully, he had a serious message that while we struggle to save old-growth forests, trees in cities where most Canadians live have also been falling to the chain saw.

Trees are a crucial link to our natural roots, reminding us of the changing cycles of life and providing shelter and food for birds, insects, mammals, and microorganisms. Of course, trees are an esthetic part of the cityscape, and if you've ever stepped from a hot concrete road into the shade of trees, you know they regulate temperature. Trees store water and transpire it into the air. They prevent erosion and do their bit to compensate for our excess production of carbon dioxide.

To be fair, some people may think of trees as pests that coat lawns and roads with dead leaves, crack underground pipes, push up pavement with their roots, and fall down during storms. But seen as a community of organisms that are quite beneficial, trees deserve to be noticed and accommodated by us.

Because they valued their "forest" and were worried that too many trees were being cut down, members of Grass-roots Albany decided to take an inventory of the trees in their community. They wanted to know how many there were, the age distribution, the species and their health, and whether trees were being planted for the future. How many of us really know this in our backyards? A proposal and modest budget (CDN $2,600) were prepared, and the four-block area to be studied was blanketed with flyers asking for cooperation. Notices were posted in local stores and on poles, stuffed into mailboxes, and delivered in person by volunteers. Response was gratifying—over 350 households representing more than 95 percent in the neighborhood agreed (most enthusiastically) to allow their trees to be examined and counted.

Forestry graduate student Marshall Buchanan was hired to do the formal work. Already involved in urban reforestation, including a project in the Rouge Valley in Metro Toronto, he was enthusiastic about the level of public knowledge, interest, and support. The only hassles he encountered were with people who thought he had come from Ontario Hydro to cut the trees down. In those few blocks, they counted more than 2,500 trees (their definition of "tree" was less stringent than the "woody plant with a single stem 2 meters [6.5 feet] above the ground" used by foresters). The citizen-initiated project provides a model that can be readily followed by any group that values the trees in its neighborhood. Initiatives like this are leading us down the road away from Rio.

Involvement is empowering, so choose a group and jump in.

A Woman in Science

SCIENCE IS A PROFESSION THAT CAN BE CHARACTERIZED AS A WHITE, upper-middle-class, primarily male preserve. As such, it is a highly competitive, macho profession in which territoriality, jealousy, and vested interest often cloud the vaunted receptiveness to new ideas. Nevertheless, women are bringing new attitudes and ideas to the preserve.

To the public, scientists seem open and receptive to unexpected data and radical ideas, which they assess objectively and rationally. But that folklore seldom holds up in reality because scientists cannot transcend their humanity. They get excited and become passionate about their work, but they can also become territorial, dogmatic, jealous, tunnel-visioned, and mean.

Lynn Margulis is a remarkable scientist who has experienced the full force of that dark, human side of science. Not only has she survived, but she continues to make waves in the scientific establishment. Outspoken, original, and fearless, the University of Massachusetts professor constantly challenges us to look at the world in new ways. Margulis became embroiled in controversy in the mid-1960s. She wondered how the earliest bacterialike cells could have evolved into more complex eukaryotes, which are defined as cells containing a nucleus, a membraned envelope enclosing chromosomes, and organelles, which are distinct structures that perform such functions as photosynthesis and energy production. All plants and animals and many microorganisms are eukaryotes.

Margulis resurrected a long-ignored idea that the organelles within eukaryotes were once free-living bacteria that long ago invaded other bacterial forms. First they were parasites, then they became symbionts, contributing services for their hosts in return for a protective environment, and finally they were fully integrated into their hosts' biological makeup as organelles.

It was a radical but scientifically testable theory. However, Margulis became a pariah among her peers for her unorthodoxy. When I first met her in the late 1970s, she painfully recounted how an application for a research grant to continue her studies had been rejected. When she called to inquire why, she was told, "Your research is shit. Don't ever bother to apply again."

But she persisted, and now many studies have shown that organelles have DNA very similar to bacterial DNA. Today, the bacterial origin of organelles is found in most textbooks, and Margulis is an eminent member of the scientific establishment. She stresses the important evolutionary role of cooperation rather than competition, pointing out that at least 10 percent of our body weight is organelles that were once separate bacteria and are now part of the cells of which we are made. Each of us is actually an immense community of organisms.

Margulis continues to explore ideas at the very edge of scientific thought. Today, she focuses on the puzzling stability of the Earth's atmosphere and ocean salinity throughout the 3.5 billion years since life began. She champions British chemist James Lovelock's proposal in 1972 that there exists some kind of self-regulation by the sum of all life-forms on Earth and their physical and chemical environment. This living skin around the planet, according to Lovelock and Margulis, is like an immense organism that has compensatory mechanisms to handle changes over time. For example, the waning intensity of the sun could have been counteracted by the production of more greenhouse gases. Too much warming could have been redressed by the release of compounds that induce clouds and cool the planet.

Lovelock named the supra-organism Gaia after the Greek goddess of the Earth, and it has captured the lay public's imagination. From a Gaian perspective, human beings are a small part of the global biosphere, and while we are changing the biological and physical properties of the planet, the survival or extinction of our species is of little consequence.

I recently talked to Margulis. As always, she was outspoken and provocative. In approaching the subject of Gaia, I suggested that we are special as the only life-forms on the planet with self-consciousness. "The dictionary," she replied, "defines consciousness as being aware of the environment. By that definition, virtually all species have consciousness." Margulis pointed out that most species of plants, animals, and microorganisms can "sense" and respond to gravity, light, temperature, chemicals, a different sex, or another species. In fact, she contends, people are far less sensitive to their environments than most other organisms, and perhaps that's why we have created such a terrible environmental mess.

Lynn Margulis should be a model for scientists. Even while studying the smallest creatures, she keeps her mind on the big picture. It doesn't matter whether or not she's right. Her real value is in stimulating us to look at ourselves and the rest of the biological world in a different way. That's science at its very best.

Young People

NO GROUP HAS MORE AT STAKE IN THE RESOLUTION OF THE GLOBAL ecocrisis than today's generation of children and youth. Young people are more receptive to new ideas, not having yet invested heavily in the status quo and therefore being able to see with greater clarity. And it is youth in whom we find the greatest ecological activism.

Try this. Take a few discarded car oil containers, store them in a heated room for a day, then pour out the residual oil. That was the science project of David Grassby, a fourteen-year-old who lives in Thornhill, Ontario. He got the idea while visiting a friend whose father was complaining about not being able to get all of the oil out of a can into his engine. David wondered how much, on average, is left when people throw "empty" containers away.

Like a good scientist, he collected more than a hundred discarded containers from trash cans and service stations. After draining one hundred of them for two minutes each, David recovered 3.7 liters (1 gallon) of oil, an average of 37 milliliters (1.25 ounces) per discard. After phoning several oil companies, he finally managed to glean enough information to calculate that annual sales of passenger car oil in Canada amount to 220 million liters (58 million gallons), of which 132 million (35 million gallons) are in 1-liter containers (executives of one company told him that is a low estimate). That means over 5 million liters (1.3 million gallons) of oil are wasted and end up contaminating soil and water annually. As well, David calculated that 10 million kilograms (22 million pounds) of empty plastic receptacles end up in dumps each year.

David then suggested that large drums of motor oil could be kept at each gas station so that motorists could fill up their own reusable container or the oil could be pumped directly into the car like gas. He sent a copy of his study to Petrocan, Shell, Sunoco, and Imperial, receiving a reply only from Petrocan. David also sent his report to the print and electronic media and the radio program *As It Happens,* which arranged for David to meet the president and executives of Esso Petroleum. At the meeting, David suggested the use of large barrels for bulk distribution of oil, but the executives replied that it was impossible because of the wide variety of grades of car oil. David replied that he had read that 90 percent of all car oil sold was 5w30. The company reps had no response.

Calling this "The Unknown Oil Spill," David printed up a brochure of his results, with suggested solutions and addresses of people to write. Like the child in the parable about the emperor with no clothes, David, with his simple science project (good scientific experiments are usually simple), went straight to the heart of a fundamental issue—unsustainable and unnecessary waste and pollution. He made people confront a number of facts: we are acting as if the environment can absorb our discards, even highly toxic ones, indefinitely; we seem to assume that our resources are so vast that we can waste them; we let the dictates of short-term profit come ahead of long-term ecological costs.

David's project also highlights the enormous cumulative impact of large numbers of tiny incremental effects. Each of us contributes a trivial amount to the planet's load, but the sum total of consumption and waste by 5.5 billion of us is enormous.

Young people like David see with embarrassing clarity because they aren't blinded by fear, vested interests in a career, or the allure of rampant consumerism. And they have the most at stake in the future of the environment. All young people today have been exposed to chemicals and toxic environmental agents from conception on, and each successive newborn will have higher exposures than any previous generation. Today's youth will become adults in a world beset with enormous ecological problems that we bequeath to them by our inability to curb the shortsighted and the unsustainable pursuit of endless growth in the economy and consumption. Their world will be radically diminished in the biological diversity that we adults took for granted when we were children.

Youth speak with a power and clarity that only innocence confers, and because we love them, adults *have* to make changes in the way we live.

Monteverde and Children

AS EVEN MORE PEOPLE BECOME AWARE OF THE GLOBAL ECOCRISIS and are convinced of the need for change, it is still difficult to find concrete things that individuals can do and that produce immediate results. There are few happy endings in the global struggle to save wilderness from the relentless pressures of poverty, overpopulation, shortsightedness, and greed. On a recent trip to Central America, I learned of a delightful story with a happy outcome. Canadians have an example that has inspired people around the world. And children played a key role.

The account begins high in the swirling mists of the Tilaran Mountains of northwestern Costa Rica. People there paid little attention to a bright orange toad that lived in the remote forest. But thirteen-year-old Jerry James, who lived there, did and reported it to Jay Savage, a visiting frog expert, in 1982. When Savage first saw the toad, its color was so extraordinary that he believed the boy had painted it. But when Jerry helped Savage collect the animals himself, the scientist was convinced and two years later, the frogs were identified as a brand-new species, *Bufo periglenes,* known locally as the "golden toad."

Because of the toad, a small area around its habitat was designated the Monteverde Cloud Forest Preserve. But like Brazil, Costa Rica allowed people to claim land simply by clearing it. As logging, burning, and mining pressed in on Monteverde, a group of landowners, farmers, and biologists realized that the preserve had to be enlarged to protect other animals, such as tapirs and the spectacular bird called the quetzal. So, in 1986, the Monteverde

Conservation League, a private, nonprofit group, was formed to buy land as a buffer around the preserve. They appealed to conservation groups in other countries and raised enough money to buy 6,559 hectares (16,207 acres) of primary forest, more than doubling the area of the preserve.

At that time, Adrian Forsythe, a Canadian naturalist, wrote an article in *Equinox,* describing the venture and suggesting that $100 would buy a hectare of cloud forest. It was a concept that environmentalists instantly knew was a great idea. In 1987, I took part in a fund-raising event for Monteverde at the University of Toronto, and the mood in the packed audience was electric as people contributed $43,000. Ontario environment minister Jim Bradley gave $10,000 on behalf of the province's citizens, a symbol of *our* stake in the forests of distant lands. The Canadian effort, supported by the World Wildlife Fund, raised over $350,000, which biologists at Monteverde assured me helped to stop a very destructive road through the preserve.

In 1987, children in a small primary school in rural Sweden began to study tropical rain forests. Nine-year-old Roland Tiensuu asked what he could do to protect them and the animals they supported. His question prompted his teacher, Eha Kern, to invite a visiting American biologist, Sharon Kinsman, to talk to the class. Kinsman had studied in Costa Rica and showed slides of Monteverde, which inspired Kern's class to raise enough money to buy 6 hectares (15 acres) of cloud forest. Kern and her husband then organized Barnens Regnskog (Children's Rainforest), through which Swedish schoolchildren raised hundreds of thousands of dollars for Monteverde.

In 1988, Sharon Kinsman formed Children's Rainforest U.S., and Tina Joliffe of England established Children's Tropical Forests U.K. In 1989, Nippon Kodomo no Jungle (Children's Rainforest Japan) was created. With the involvement of children on an international scale, the Conservation League decided to buy forest specifically for children. Today, Bosque Eterno de los Niños (Children's Eternal Forest) has almost 7,000 hectares (17,300 acres), with a long-term goal of over 16,000 hectares (39,500 acres). In 1990 alone, children raised over a million dollars for the International Children's Forest at Monteverde.

Currently, more land is still being purchased around Monteverde at $250 per hectare. But it is just as important to provide money to protect the land that is already in the preserve. Peasants around the preserve must be educated about the value of wilderness to reduce poaching, illegal logging, and squatters. Farmers are being taught to reforest their land, primarily as windbreaks that protect pasture and increase productivity of land already cultivated. The planted trees provide fence posts, lumber, and firewood that would otherwise come out of the forest. The education program and tree nursery need funds. More guards and a rapid telephone system are needed. The staff of the Children's Forest dream of building an education center with an amphitheater for young people from around the world.

Canadians continue to support the preserve and can be proud of their role in Monteverde.

Update

The Children's Eternal Rainforest is located on the pacific slope in a transition zone between premountain wet and moist forests, so it is extremely rich in biodiversity. Jaguar Canyon, which runs through a small part of the park, was found to contain thirty tree species that were new to science. The park is home to mammals such as the sloth, white-faced monkey, agouti, coatimundi, kinkajou, marguay, porcupine, hog-nosed skunk, fox, coyote, armadillo, and bat. Now encompassing more than 20,000 hectares (49,000 acres), the Children's Rainforest is the largest private reserve in Central America and is administered by the nonprofit Monteverde Conservation League.

Climate is affecting the cloud forest. As the dry season has become warmer and longer, the permanent cloud cover is moving up the mountains. The golden toad that catalyzed the formation of the park in the first place already lived at the very tops of the mountains. In 1987 and then from 1991 to 1994, Dr. Alan Pounds examined a 30-square-kilometer (12-square-mile) area of Monteverde and found that twenty of fifty known species, including *Bufo periglenes,* the golden toad, were not to be found. He believes climate change is the cause.

Child Power

"AND A LITTLE CHILD SHALL LEAD THEM" (ISAIAH 11:6).

In the Old Testament, it is only after Armageddon that children become leaders. Let's hope it doesn't take an environmental Apocalypse before we listen to our children. Like the child in the parable about the emperor with no clothes, most children can see with clarity and innocence and tell it like it is.

We grown-ups say that our children matter to us more than anything else on Earth. If so, that love should make us care deeply about the kind of world they will grow up in. Over the half-century of my lifetime, the planet has changed beyond belief—the once-vast assemblages of wildlife and seemingly endless ancient forests have been drastically reduced to mere vestiges of what they were. We know with absolute certainty that our children will inherit a world with radically diminished biological diversity and extensive global pollution of air, water, and soil. If we do love our children, what excuse can we possibly have for not pulling out all stops to try to ensure that things don't get worse?

Children may be powerless politically, economically, and legally, yet they are the ones with the greatest stake in the decisions that are or are not being made right now. Most of the people currently in power in government and business will not have to live with the consequences of their action or inaction; it will be today's youngsters who become adults in the twenty-first century. That's why children have to take an active role in shaping their own destiny.

One way is to influence their parents. Schools are having an enormous impact on children, making them more aware of environmental issues than many of their parents. Those parents are important people—lawyers, laborers, doctors, homemakers, politicians—that is, the people who make up all of society. Those environmentally concerned children, then, must affect the most important adults in society, because if they can't, who can?

My daughter, Severn Cullis-Suzuki, is now a thirteen-year-old in her second year of high school and is proudest of making the basketball team. She has accompanied Tara and me to many demonstrations, peace marches, and environmental events since infancy, so activism has come to her naturally. I remember finding Severn on the sidewalk when she was seven, selling hardcover books taken from our shelves for 25 cents so that she could send money to support Chief Ruby Dunstan's fight to save the Stein Valley in British Columbia.

In 1990, I took my family to live with Kaiapo people in the middle of the Amazon rain forest in Brazil. It was like stepping back five thousand years in time. Severn, then ten, had a wonderful time with newfound Kaiapo friends. When we left, it was with great regret, all the more so because as we flew out of the forest, we could see gold mines and fires devastating Kaiapo land. Out of fear for her Kaiapo friends, Severn talked a group of Grade 5 girls into forming a small group that she called the Environmental Children's Organization (ECO) to spread the word about environmental issues. ECO started making ceramic brooches shaped like geckoes. Calling the products eco-geckoes, the girls displayed them in school and were deluged with orders from students and teachers alike.

Soon ECO had raised over $150. Inspired by a speech given by Thom Henley about the plight of the Penan people in Sarawak whose forest home was being destroyed by logging, the club bought a large water filter and gave it to two Penan people who visited Vancouver. The girls gave slide shows, talked at schools and youth meetings, and gained a profile locally. In the summer of 1991, Severn told me she had heard about "a big environmental meeting" in Rio and asked whether I was going. I told her no and was surprised when she said, "I want to raise money to send ECO. I think children should be there to act as a conscience for the grown-ups." I scoffed at the idea, warning her it would be dangerous, polluted, frustrating, and expensive, then promptly forgot about it. Two months later, Severn walked in and announced, "Dad, I just got a check for $1,000 from the Ira-hiti Foundation in San Francisco." Only then did I learn that when Doug Tompkins, the founder of the Ira-hiti Foundation, had visited Tara and me during the summer, he had talked to Severn and she had told him about her idea. He had encouraged her to apply for support. When she did, he came through.

I had not expected Rio to accomplish much, but seeing how serious Severn was and reflecting on her idea, I realized that children like her might be able to say things adults couldn't or wouldn't. So I told her if she was serious, I would pay for her way, as well as for Tara and me as chaperones, if she and her club could raise the rest. They raised $13,000! That was enough with my matching funds for five girls and three parents to go to Rio.

Once we registered for the Rio meetings, both Tara and I were asked to take part in various programs of the Earth Summit, the Global Forum, and the Earth Parliament. At each event, we used the occasion to let the girls give brief talks. They were a sensation and received standing ovations while many in the audience wept openly.

After Severn and Michelle Quigg had given talks in Tara's program at the Earth Parliament, William Grant, the American head of UNICEF, who was in the audience, rushed up and asked Severn for a copy of her speech. "I will personally give this to Mr. Mulroney when he arrives," he promised her. That night, we learned that Grant had run into Maurice Strong and urged him to let Severn speak at the Earth Summit. In violation of protocol, Strong put Severn on the program with three other girls who were representing official youth organizations.

Severn worked hard on writing the speech, politely rejecting my suggestions: "Daddy, I know *what* I want to say. I want you to teach me *how* to say it." The speech had an impact that continues to reverberate. As soon as it was over, then-Tennessee senator Al Gore rushed over and told her, "That was the best speech given at Rio." In his closing remarks, Maurice Strong ended by quoting from Severn's talk. The official UN video of the Earth Summit concludes with clips from Severn's talk. The video of her speech has received wide exposure around the globe and provides an example that a determined individual, even if a child, can have an impact.

Here is the speech she gave on June 11, 1992, at a Plenary Session of the Earth Summit at RioCentro, Brazil:

> Hello, I'm Severn Suzuki, speaking for ECO, the Environmental Children's Organization.
>
> We are a group of four twelve- and thirteen-year-olds from Canada trying to make a difference—Vanessa Suttie, Morgan Geisler, Michelle Quigg, and me.
>
> We raised all the money ourselves to come 6,000 miles to tell you adults you *must* change your ways.
>
> Coming up here today, I have no hidden agenda. I am fighting for my future.
>
> Losing my future is not like losing an election or a few points on the stock market.
>
> I am here to speak for all future generations yet to come.
>
> I am here to speak on behalf of the starving children around the world whose cries go unheard.
>
> I am here to speak for the countless animals dying across this planet because they have nowhere left to go.
>
> I am afraid to go out in the sun now because of the holes in the ozone.
>
> I am afraid to breathe the air because I don't know what chemicals are in it.
>
> I used to go fishing in Vancouver, my hometown, with my dad, until just a few years ago we found the fish full of cancers.
>
> And now we hear about animals and plants going extinct every day—vanishing forever.
>
> In my life, I have dreamt of seeing the great herds of wild animals, jungles, and rain forests full of birds and butterflies, but now I wonder if they will even exist for my children to see.
>
> Did you worry about these things when you were my age?
>
> All this is happening before our eyes, and yet we act as if we have all the time we want and all the solutions.
>
> I'm only a child and I don't have all the solutions, but I want you to realize, neither do you!

You don't know how to fix the holes in our ozone layer.

You don't know how to bring the salmon back up a dead stream.

You don't know how to bring back an animal now extinct.

And you can't bring back the forests that once grew where there is now a desert.

If you don't know how to fix it, please stop breaking it!

Here you may be delegates of your governments, business-people, organizers, reporters, or politicians. But really you are mothers and fathers, sisters and brothers, aunts and uncles. And all of you are somebody's child.

I'm only a child, yet I know we are all part of a family, five billion strong—in fact, 30 million species strong. And borders and governments will never change that.

I'm only a child, yet I know we are all in this together and should act as one single world toward one single goal.

In my anger, I am not blind, and in my fear, I'm not afraid to tell the world how I feel.

In my country, we make so much waste. We buy and throw away, buy and throw away. And yet northern countries will not share with the needy. Even when we have more than enough, we are afraid to lose some of our wealth, afraid to let go.

In Canada, we live the privileged life with plenty of food, water, and shelter. We have watches, bicycles, computers, and television sets.

Two days ago here in Brazil, we were shocked when we spent time with some children living on the streets.

And this is what one child told us: "I wish I was rich. And if I were, I would give all the street children food, clothes, medicine, shelter, love, and affection."

If a child on the street who has nothing is willing to share, why are we who have everything still so greedy?

I can't stop thinking that these children are my own age, that it makes a tremendous difference where you are born. I could be one of those children living in the *favelas* of Rio. I could be a child starving in Somalia, a victim of war in the Middle East, or a beggar in India.

I'm only a child, yet I know if all the money spent on *war* was spent on ending poverty and finding environmental answers, what a wonderful place this Earth would be.

At school, even in kindergarten, you teach us how to behave in the world.

You teach us

- not to fight with others;
- to work things out;
- to respect others;
- to clean up our mess;
- not to hurt other creatures;
- to share, not be greedy.

Then why do you go out and do the things you tell us not to do?

Do not forget why you are attending these conferences, who you are doing this for—we are your own children.

You are deciding what kind of a world we will grow up in.

Parents should be able to comfort their children by saying, "Everything's going to be all right." "We're doing the best we can." "It's not the end of the world."

But I don't think you can say that to us anymore.

Are we even on your list of priorities?

My dad always says, "You are what you *do,* not what you *say.*"

Well, what you do makes me cry at night.

You grown-ups say you love us. I challenge you, *please,* make your actions reflect your words. Thank you for listening.

Update

Severn's speech had an amazing impact, which in turn influenced the course of her life. She was interviewed extensively by the media in both

Canada and the United States. Elected to the United Nations Environmental Program's Global 500 Roll of Honour, she went to Beijing, China, to be inducted. She wrote a best-selling book, *Tell the World,* which described her experiences and included the text of her Rio speech. Severn was elected to a committee to draft the Earth Charter, headed by Maurice Strong and Mikhail Gorbachev and chaired by Stephen Rockefeller.

Severn's proudest achievement was being a starting player on her high school basketball team. In 2000, she bicycled across Canada for clean air. She majored in Ecology and Evolution at Yale University, graduating in 2002. She formed an environmental think tank called the Skyfish Project, which seeks to take responsibility for the future by implementing concrete lifestyle changes that lighten one's impact on the environment. As a member of UN Secretary General Kofi Annan's committee of eminent persons, Severn helped outline the objectives of the Earth Summit in Johannesburg in 2002.

Germany—An Inspiring Example

INDUSTRIALIZED COUNTRIES, INCLUDING THE UNITED STATES AND Japan, have been living through a deep and prolonged recession. In times of government restraint, as politicians and businesspeople try every means to prime the economic engine, it is assumed that environmental issues must take a back seat. So it is heartening to learn that one country that has been a global economic powerhouse does not see the matter in the same way.

In 1988, delegates at an international conference on atmosphere change held in Toronto called on governments to reduce carbon dioxide emissions by 20 percent in fifteen years. On its own initiative, West Germany announced an even more stringent target of 25 to 30 percent reduction by 2005. (By contrast, the Atmosphere Convention adopted at the Earth Summit in Rio and signed by Canada seeks merely to stabilize carbon dioxide emissions at 1990 levels by 2000.) West Germany set its goal before it was reunited with East Germany, whose industry is antiquated and polluting. Nevertheless, the country is still striving to meet its original target.

For years, industrialized countries have promised to commit 0.7 percent of their GNP to aid poor nations. At Rio, rich countries reluctantly renewed that promise, but only Holland and the Scandinavian countries actually meet the target. Even though the costs of reunification and cleaning up East Germany's industry have been staggering, at Rio, Chancellor Helmut Kohl vowed to make a serious attempt to meet the 0.7 percent commitment "as soon as possible."

Since the ozone layer has been depleted faster than expected, the Montreal Protocol target of eliminating CFCs in industrialized countries by 2000 was advanced by two years. Canada hopes to eliminate CFCs even sooner, while Germany announced plans to eliminate CFCs by 1995 at the latest and is trying to phase them out by 1993! [And they did.]

These are impressive environmental commitments. Recently, I spoke with Dr. Klaus Schmidt, first secretary of the German embassy in Ottawa, who told me German politicians and businesspeople take it for granted that the health of the economy cannot be separated from the quality of the environment. As evidence, he ticked off government and industry initiatives. For example, the Steilmann Institute has developed a line of clothing made completely from natural fibers and colors without the use of any added chemicals.

In anticipation of federal laws, the German auto industry is designing cars so that their components can be removed and reused. Auto designs now emphasize reduced environmental impact and safety (Mercedes was the first to have air bags as a standard feature).

German law requires that packaging that is not an essential component of a product must be taken back by the merchant. The merchant may reuse the packaging or return it to the manufacturer, who cannot just dump it into the waste stream but must recycle or reuse it.

Most industries in North America, as in other countries of the European Community, are not prepared to emulate these ecologically sensible practices. Instead, they often demand greater freedom from government regulation to enhance their global competitiveness. As in Japan, German industry works closely with the government and trade unions in a kind of economic and environmental partnership. It is assumed the added costs of being environmentally responsible will be recovered in the long run, but Schmidt declared, "We have to do this, anyway, because it is the right thing to do."

Why is Germany, rather than Britain, France, or Italy, an environmental leader? For one thing, German culture has a rich mythology about nature. For another, the country is already paying the ecological costs of industrial development. The famous Black Forest is dying as a result of acid rain created by European industry. German "forests" are really third- and fourth-generation tree plantations that lack the rich diversity of real forests and are highly vulnerable to environmental stress.

Germans have long agonized over their location between the nuclear superpowers. Chernobyl set off near hysteria over nuclear fallout. When ten thousand seals died mysteriously in the North Sea, Germans knew the animals were a biological warning. And the chemical spill at Basel poisoned the Rhine and caused a massive kill of plants and animals in a river that is deeply embedded in German folklore.

I have often met German tourists in remote parts of Canada, reveling in the beauty of nature that can't be matched in their own country. They often scold us, telling us we should value and care for our natural wonders far more than we do. And modern Germany shows that being ecologically responsible makes economic sense.

Update

Germany, like other European nations, such as Denmark, Norway, Holland, and Sweden, is leading the world in recycling, energy efficiency, and alternative energy. Several examples of positive steps toward sustainability in Germany are cited in my book *Good News for a Change: Hope for a Troubled Planet*. In the area of wind power, Germany has become the world's leading exporter of wind technology. The country itself has about half of all the turbines in existence, more than eleven thousand by 2002. In 2001, the number of turbines in use rose by 44 percent, providing 3.5 percent of all power for the country. The dramatic expansion occurred after the government decided to phase out all nuclear power.

Germany is planning to build immense wind parks containing about five thousand huge turbines up to 45 kilometers (28 miles) offshore. The European wind power industry estimates that with proper legal and financial support, wind could provide energy for 50 million Europeans in less than ten years.

Water and a Canadian Scientist

BEING SO CLOSE TO THE UNITED STATES, CANADIANS OFTEN FIND IT difficult to point to genuine homegrown heroes. This is especially true in science, where indigenous experts are often tempted by salaries, prestige, and research support to leave. It is therefore inspiring to find a story that is both scientifically and environmentally uplifting.

It begins in 1964 when the International Joint Commission recommended a study of the state of the water in the Great Lakes. A year later, an interim report documented significant eutrophication, an explosive growth of algae due to overfertilization that ultimately sucks up oxygen and chokes off other life-forms. The study urged more research to identify the impact of human activity on water and to develop guidelines for legislation.

In response, the Fisheries Research Board of Canada, a government unit now absorbed by the Department of Fisheries and Oceans, began to seek a research area. Forestry companies and the Ontario government agreed to set aside a tract of forest southeast of Kenora, Ontario, for research only. The Experimental Lakes Area (ELA) contained forty-six lakes, ranging from 5 to 60 hectares (12 to 148 acres), on which experiments could be performed.

In 1968, ecologist David Schindler was recruited to spearhead the studies of the ELA. Under his direction, the ELA has proved to be a scientific mother lode, yielding results recognized by scientists around the world and affecting legislation in Canada, the United States, and the European Community. The first research project was prompted by severe eutrophication in Lake Erie. By adding controlled amounts of carbon, nitrogen, and phosphorus compounds to different lakes, Schindler's team confirmed in 1973 that the major contributor to eutrophication was phosphates from sewage and detergents (where they "put brightness in your wash"). Canada immediately limited phosphates in detergents and urged more tertiary sewage treatment. In the United States, Schindler influenced many states to follow suit in the ensuing years, although Ohio didn't act until 1988. As phosphate levels fell, the recovery of Lake Erie and lakes in Haliburton and Muskoka in Ontario was spectacular and corroborated the value of the ELA.

Early on, Schindler recognized acid rain as a concern to come and, in 1974, before the government developed a policy, began to study the lakes so that the deliberate addition of sulfuric acid could be started in 1976. The studies showed that as a lake became more acidic, plant and animal species disappeared in a specific sequence. As prey species at the base of the food chain were lost, there was "a cascade of extinction" back up to species of large fishes. When acidification was stopped and the waters diluted out the acid, species began to return in the reverse order from their disappearance (the recruits came back through rivers connecting the lakes). Through these studies, Schindler was able to set levels of acidity that could be absorbed and still allow the recovery of lakes. Those levels became targets to be set by governments and negotiated in transnational discussions. The ELA work on eutrophication and acidification has resulted in over CDN $8 billion in government commitments to implement the standards that resulted.

Over the twenty-year span of the ELA, the average temperature in the region has risen more than two degrees Celsius. In what could be his most significant work, Schindler and his colleagues have begun to measure the effects of temperature on the biological makeup of the lakes as a prelude to understanding the long-term consequences of global warming.

The announcement of the Stockholm Water Prize, a new award comparable to a Nobel Prize, says, "One of the greatest threats to our very existence today is environmental pollution, not least of our water. Water is such a prerequisite for all life on this Earth that it cannot be a matter of only national concern: it is an international question." On August 14, 1991, David Schindler was awarded the First Stockholm Water Prize of U.S. $150,000.

Schindler richly deserves the recognition, and all Canadians should take pride in the farsighted commitment made by federal and provincial governments and the team of scientists who participated in this outstanding project.

Update

David Schindler's work is recognized around the world. He is the only recipient of both the Volvo Prize and the Stockholm Water Prize. On November 5, 2001, Schindler earned the highest prize in Canada, the Gerhard Herzberg Gold Medal for Science and Engineering, which awards $1 million for research.

Schindler richly deserves his awards. He continues to be a powerful research scientist, investigating the "grasshopper effect" that deposits volatile compounds that precipitate out when the upper atmosphere encounters cold fronts. He constantly reminds politicians, businesspeople, and the public that the atmosphere is a shared commons and that we have to design better ways to interact with it. He is uncompromising in his statements on issues such as climate change, because his views rest on the best scientific evidence.

One Logger and His Forest

IN MANY PARTS OF THE WORLD, THE MOST CONTENTIOUS ECOLOGICAL disputes rage over the future of forests. In European countries such as Germany and Sweden, experience indicates that trees cannot be repeatedly grown and harvested like agricultural crops. Sustainable logging must always maintain the complexity and integrity of the forest ecosystem. Doing so requires careful and selective logging practices, but it can be done.

There are a few articulate and persuasive foresters, such as Oregon's Chris Maser, who argue that forestry practices must be radically changed. Merv Wilkinson is another. Wilkinson's thoughts on forests are recorded by Ruth Loomis in *Wildwood: A Forest for the Future.* This little book is studded with nuggets of common sense and wisdom that expose how short-sighted current large-scale forestry practices are. Wildwood is Wilkinson's 55-hectare (136-acre) lot on Vancouver Island, which he has owned since 1939. In the fifty-one years of his tenure, he has logged it nine times for a third of his income (it took 20 percent of his working time). In 1939, it was estimated that Wildwood contained about 1.5 million board feet of usable timber. By 1992, another 1.5 million board feet will have grown in the same forest. In the same interval, Wilkinson has logged almost 1.4 million board feet. In other words, even though the equivalent of almost the entire original forest has been cut over five decades, there is still as much wood as there ever was! Equally important is that the forest ecosystem has remained *intact* throughout the time. Wilkinson calculates that only 200 to 240 hectares (500 to 600 acres) cared for like Wildwood would keep two people fully employed as well as a crew aud trucker at falling time *and* the forest would never disappear.

Wilkinson says: "The essential ingredient in effective woodlot management is time—a long-term perspective and a day-to-day participation in a living landscape that evolves over decades and even centuries." His operating rule is simple: "Work with nature! Invariably 'nature knows best' and my instinct says nature's ways are vastly superior to human ways." So he logs selectively, minimizes road construction, and always works within the constraints imposed by trees, soils, and topography themselves. Chemicals were never used in Wildwood, yet a spruce budworm outbreak in 1939 has never recurred and the incidence of diseases such as "conk" and "root rot" has declined.

Wilkinson's ideas also make *economic* sense:

> I never cut over the annual growth rate ... [I] consider the forest the "bank account," the annual growth the "interest." The "interest" is converted into the products which are removed from the forest, but the "account" is left standing. I have now learned to leave five percent of my "interest" or annual growth to decay and rot on the forest floor, a reinvestment in the soil, a reinvestment for the future ... in British Columbia. The big companies have been allowed to over-cut, abandoning any idea of sustaining. The Department of Forestry claims that in 1988 the forests grew 74 million board feet during the year. The cut during that year was 90 million.... Is there any worse "deficit financing" than that?

To Wilkinson, a forest is far more than trees. It "includes the soil, with all its interdependent bugs, fungi, burrowing mammals, ground-covers and undergrowth, the trees themselves, the birds and animals living in or moving through it, the natural water systems and the air. The basic cornerstones of forestry are soil, water, air, and sunlight. One cannot be separated from the other." Large-scale logging and replanting focus on the trees as if they are all there is. But Wilkinson knows the forest "is a balanced entity, so that if you destroy that balance, you're going to be in trouble."

Wildwood is proof that it is possible to derive a living from a small forest while maintaining it as a diverse ecosystem. In contrast with the "cornrow, industrial-agricultural style of tree farming," Wilkinson's lot "has been both profitable and aesthetically intact ... there are still abundant populations of native wildlife; eagles, pileated woodpeckers, owls, deer, and a multitude of other creatures in their own habitat. The traditional methods of forestry have been primarily interested in reestablishing trees but not the complex ecosystem of a forest."

Wilkinson warns: "So far, management of our forests has been in the hands of those who do not recognize the forest as an ecosystem of all ages and species which are interdependent.... Boardroom foresters cut for the product without respect for the life of the forest." The battlelines over the future of Canada's forests revolve around those differences in perspective. It's time the government paid more attention to people like Merv Wilkinson.

Philosopher-King

NINETEEN NINETY-TWO WAS A TOUGH YEAR FOR THE BRITISH ROYAL Family. *Annus horribilis,* Queen Elizabeth called it. In 1993 the press began reporting the seamy details of the demise of the Prince and Princess of Wales's marriage. Prince Charles, the king long-in-waiting, has been portrayed by the British tabloids as a lout with big ears, a kook who talks to plants, an uninformed meddler in the field of architecture, and so on. But anyone who has read his writings or listened to his speeches will find a man who is highly informed, deeply concerned about the state of the planet, and actively involved in trying to do something to make a difference.

Consider the remarkable speech he gave at Kew Garden in London on February 6, 1990. It is a perceptive and moving account of his concern about the future of the world's tropical rain forests. Here is some of what he said:

> For hundreds of years the industrialized nations of the world have exploited, some would say plundered, the tropical forests for their natural wealth. The time has now come to put something back, and quickly.... The forests assist in the regulation of local climate patterns, protecting watersheds, preventing floods, controlling huge flows of life-giving water. As the forests come crashing down, an inexorable human tragedy is set in train.... The whole of humanity will benefit if what is left of the tropical forests can be saved.

He then described the contribution to global warming of burning as well as loss of the carbon-removing potential by deforestation. He also pointed to the very real potential of forest species to provide drugs and potential food crops: "The genetic reservoir of plant and animal life provides us with the most perfect survival kit imaginable as we face the unknown challenges of the future." The prince quotes the tropical biologist Norman Myers: "Tropical forests have lost 142,000 square kilometres [55,000 square miles] of their expanse during 1989. This is 1.8 percent of remaining forests."

His Royal Highness clearly understands that "the main cause [of forest destruction] is the poverty of people who live around the forests, together with the inexorable pressure of ever-growing human numbers.... The time has come for an international agreement or convention on the world's tropical forests." The goals of a Rainforest Convention, he suggests, would include the development of sustainable use of forests, maintenance of maximal biodiversity, protection of rights of aboriginal forest dwellers, beginning reforestation, compensation for lost revenue, and establishment of funding mechanisms.

The Prince of Wales points out that a major destructive agent is the international debt. "In 1989 the South paid $52 billion more to the North in the way of debt servicing than it received in the form of foreign aid.... Once the forests are thought to hold a greater hope for human development and economic development if *conserved,* then it clearly becomes possible to reconcile environmental protection and development."

His speech was very careful to point out the importance of respecting both the rights of indigenous forest people and their vast knowledge of the ecosystem. "Generations of observation and bodily trial and error have honed their judgment in a process as rigorous as any laboratory testing.... These people are accomplished environmental scientists and for *us* to call *them* primitive is both perverse and patronizing." He went on to think of the forests themselves: "There are thousands fewer tropical forest trees than there were when I started speaking and they can't speak for themselves. They have a voice of their own, but it's only a whisper and hard to hear above the shriek of the saw."

The Prince stressed that we in the industrialized world must stop using tropical hardwoods that come from old-growth tropical rain forests, help countries use forests sustainably, and relieve the crippling burden of international debt. He ended his speech this way:

> I fear that we will fail this challenge if we are not prepared to accept that sustainable development demands not just a range of different management techniques and funding mechanisms, but a different *attitude* to the Earth and a less arrogant, man-centred philosophy. We need to develop a reverence for the natural world. The tropical forests are the final frontier for humankind in more ways than one. Our efforts to protect them will not only determine the quality of life and economic security of future generations, but will test to the limit our readiness to cast off the kind of arrogance that has caused such devastating damage to the global environment, and to become the genuine stewards of *all* life on Earth, not just the human bit of it.

These are thoughtful, even revolutionary words. Coming from the possible future king of England, the words have a powerful impact. Charles seems on his way to becoming a genuine philosopher-king, an inspiration and leader we so desperately need.

A New Kind of Political Leader

THE TURBULENT ERA OF THE 1960S AND 1970S ENDED WITH THE retirement of Pierre Trudeau and the end of the Vietnam War. Nineteen ninety-three signaled the passing of the era of globalization of the economy and social meanness epitomized by U.S. president Ronald Reagan and Prime Ministers Margaret Thatcher of Britain and Brian Mulroney of Canada. No one illustrated the lack of genuine concern about the environment more than George Bush. Cloaking himself with green rhetoric, Bush declared himself a future "environmental president" while castigating the ecological record of his opponent, Massachusetts governor Michael Dukakis. Once elected, Bush quickly revealed his contempt for environmentalists and became a symbol of eco-ignorance by his crude bullying of delegates preparing for the Earth Summit in Rio. In the dying days of the 1992 election, Bush mocked Al Gore, the Democratic vice presidential candidate, as the "Ozone Man." So the victory of the Clinton–Gore team provided a much needed boost to the millions of people concerned about the state of the Earth.

I remember in high school in London, Ontario, in the early 1950s, when a fellow student told me he hoped to go into politics. He was a school leader with high ideals to make Canada something better, and we all admired and envied him. Years later, I thought of him often and always encouraged our "best and brightest" to consider a career in politics because we need people in government with vision, courage, and integrity. Unfortunately for many today, a political career costs too much in income, ethics, and self-esteem.

Meanwhile, polls tell us public respect for politicians is plummeting. Perhaps federal government actions on the environment reveal the reasons for the erosion of public confidence.

In 1989, I interviewed a politician who gave me goose bumps as he recited the dimensions and severity of the global ecocrisis and his commitment to doing something about it. He was Al Gore, then the U.S. senator from Tennessee who had run in 1988 for the Democratic nomination for president with the environment as his top priority. Back then the American media declared that the environment was not a "presidential issue," and Gore lost badly. The senator took time to reflect on that loss and, as a result, wrote the best-selling book *Earth in Balance: Ecology and the Human Spirit,* because, he says, "I cannot stand the thought of leaving my children with a degraded Earth and a diminished future."

The book is frightening, profound, and ultimately inspiring as it takes an unblinking look at the dimensions of the crisis, points out the root causes, and then offers a concrete and detailed strategy to avoid a total collapse in the ecosphere. Gore portrays a planet being ravaged by human numbers, consumption, and technology. The familiar litany of ecological facts still shocks—1.7 billion people lack access to clean water, 25,000 people die from waterborne diseases daily, world chemical production doubles in volume every seven to eight years, pesticides are made "today at a rate 13,000 times faster" than in 1962, "every person in the United States produces more than his or her weight in waste every day."

Gore's personal analysis and convictions give the book its power: "We are creating a world that is hostile to wilderness, that seems to prefer concrete to natural landscapes." To Gore, the planetary crisis "is an outer manifestation of an inner crisis that is, for lack of a better word, spiritual." We have lost a sense of wonder and awe once inspired by a feeling of belonging and kinship with the rest of the living world. Thus, Gore suggests, we are a dysfunctional species that compensates for our alienation from nature by overconsumption. We are no longer rational because "civilization is, in effect, addicted to the consumption of the Earth itself."

Gore indicts politicians, including himself, for getting caught up in the superficiality of politics: "Voice modulation, 10-second 'sound bites,' catchy slogans, quotable quotes, newsworthy angles, interest group buzzwords, priorities copied from pollsters' reports.... The environment is not just another issue to be used in political games for popularity, votes, or attention. And the time has long since come to take more political risks—and endure much more political criticism—by proposing tougher, more effective solutions and fighting hard for their enactment."

He attacks economics for greatly distorting our relationship to the world: "In calculating GNP, natural resources are not depreciated as they are used up." Economists make "absurd assumptions that natural resources are limitless 'free goods' " because it leads us "to act as if it is perfectly all right to use up as many natural resources in our own lifetime as we possibly can." These are rare, courageous, and perceptive insights from a politician.

In the end, Gore finds in Christianity as in other great religions, lessons about our responsibility to care for God's creations so that we can pass what we have inherited to future generations. He asks: "If the Earth is the Lord's and we are given the responsibility to care for it, then how are Christians to respond to the global vandalism now wreaking such unprecedented destruction on the Earth?"

His overriding concern is for future generations as we fail "to look beyond ourselves to see the effect of our actions today on our children and grandchildren.... We care far less about what happens to our children than about avoiding the inconvenience and discomfort of paying our own bills.... As we strip-mine the Earth at a completely unsustainable rate, we are making it impossible for our children's children to have a standard of living even remotely similar to ours."

Gore skewers politicians' tendency to demand more information because it "is actually an effort to avoid facing the awful, uncomfortable truth: that we must act boldly, decisively, comprehensively, and quickly, even before we know every last detail about the crisis." His solution is a "global Marshall Plan" to save the world. He lists five strategic goals: stabilize world population, develop appropriate technologies, formulate an ecological economics, make international agreements on the environment, and raise people's awareness around the world. Each goal is chosen with care, and strategies for achieving them are presented in considerable detail (including strong recommendations for the American role). This carefully thought-out book portends a new generation of politicians with a holistic vision for the Ecological Millennium. Every politician, public servant, businessperson, and economist should read it. Gore gives proof that is possible to be a politician and still uphold the kind of ideals of that high school student I envied so long ago.

Update

In 1989, Al Gore was visiting Canada when I was preparing for a CBC radio series, *It's a Matter of Survival.* I interviewed Gore for the series and have never met another politician who could articulate so clearly the ecological crisis that confronts us. His book confirmed that he knew what the issues were.

When I turned off the recorder, I commented facetiously, "Please immigrate to Canada and I will do all I can to get you elected Prime Minister." In a more serious vein, I asked him, "What can people like me do to help politicians like you?" His answer changed my focus. He said, "Don't look to politicians like me. If you want change, you've got to take it to the people. Convince them there is a problem. Show them there are alternatives. And get them to care enough to demand that something be done. When you've done that," he told me, "every politician on all sides of the political spectrum will jump on the bandwagon."

I took his advice to heart, and when I formed the David Suzuki Foundation, it was to research for solutions to our environmental problems and then to communicate both the problems and the solutions to the general public. I saw Gore at the Earth Summit in 1992 and at the Kyoto meeting in 1997, where he was a hero to the delegates attending. But when he ran for the presidency in 2000, I was struck by how he deliberately avoided serious discussion about the environment, which he had been so passionate about as senator and vice president. It was then that I realized how prescient his advice to me had been. He couldn't campaign on the environment as a political issue because the American electorate wasn't ready for it.

Fisheries That Flourish

SALMON HAVE AN EXQUISITELY COMPLEX LIFE CYCLE, BEGINNING and ending in freshwater rivers, with years of foraging ocean waters in between. In attempting to manage these wild animals, we impose our will on them through the different perspectives and priorities of our economic, political, and social categories. In the process, we ensure that the salmon will not be dealt with as a single biological entity.

Fish on the Line, a report prepared for the David Suzuki Foundation (DSF) by the renowned fish biologist Dr. Carl Walters, boldly stated that management by the Department of Fisheries and Oceans is setting up a biological disaster driven by political and economic pressures. The failure of the sockeye run in 1995 appears to support Walters's prescience. But what else can we do? Are there alternatives?

The answer is yes. *Fisheries That Work* was prepared for the DSF by Dr. Evelyn Pinkerton, a maritime anthropologist who has a long involvement with fishing communities along the Pacific Coast, and Dr. Martin Weinstein, who specializes in the socioeconomic aspects of natural resource management. In case studies of fisheries around the world, they found dozens of encouraging examples of sustainable practices and focused on ten cases, in areas ranging from Peru to Australia, Japan, the U.S., Korea, and Canada.

Management by fishing villages has worked for hundreds, if not thousands, of years. Lake Titicaca in Peru has a stable fishery employing three thousand fishers who harvest more than 8,000 tonnes annually. With no financial or government support, the Titicaca fishers, from communities spanning two countries, make and enforce the rules. According to Pinkerton and Weinstein, "Perhaps the most astounding fact about this fishery is that harvest levels appear to have remained stable since the 16th century when some Spanish records are available."

Two success stories in salmon management are described in B.C. and two in Alaska. Two involve examples of cooperation between Native and non-Native fishers. In Alaska, a successful initiative was begun in the mid-1970s, when the catch had dropped sharply to 30 million salmon. As a result of state and community efforts, the catch rose to a spectacular record of 194 million salmon in 1994!

In B.C., the Skeena River Watershed Committee brings together sport, commercial, aboriginal, federal, and provincial interests to coordinate the various fisheries and to halt the decline in biodiversity of the fish. The committee shows that given responsibility and accountability, members will cooperate and can be effective.

Japan's inshore fishery of nonmigratory species has been dramatically successful. Dominated by small-scale operations, it contributes almost a third of Japan's 11-million-tonne national catch and almost 50 percent of its value, yielding profits of more than $1.5 billion in 1988. The basic fishery is more than a thousand years old and has 22,000 local cooperative associations, each with an average of 250 members. The major thrust of this system is to funnel a maximum of the benefits of the fishery back into local communities.

From the case studies surveyed in *Fisheries That Work,* three elements emerge that are common to these successful management strategies and that provide useful lessons for Canada's fisheries.

First, planning units smaller than the whole coast are needed if the project is to be effective in dealing with conflicts between different interest groups. People brought together with shared knowledge about the fish and concerns about their future can resolve conflicts, whereas solutions imposed from "higher" or central offices are resisted.

Second, agreements or contracts between government agencies and local or regional bodies allow a beneficial division of labor. Thus, the government excels at setting general goals and standards on which it holds the line, while fishers and local bodies figure out the most practical ways to implement the goals and give themselves more fishing time. Their plans tend to be more flexible, creative, and better adapted to the local situation.

Third, local or regional committees bring specific talents and resources to the process, from knowledge of fish stocks to greater trust of local communities to access to volunteer labor. Such a committee provides more comprehensive and effective management of harvest, enforcement, enhancement, and habitat protection and restoration.

The crisis in salmon demands that we get beyond the finger-pointing and look to solid success stories from which to draw inspiration and guidance. *Fisheries That Work* is a good start.

Update

The Federal Department of Fisheries and Oceans' catastrophic management of northern cod on the East Coast has been matched by its equally inept approach to Pacific salmon. Perhaps the coastal symmetry is fitting, with the DFO's offices located in the center of the country. DFO minister Fred Mifflin spent millions to reduce the fishing fleet by buying and retiring commercial licenses. However, the West Coast is divided into three fishing zones, and a license is required to fish in each zone. Instead of encouraging local fishers by restricting fishing to the zone of a fisher's residence, Mifflin allowed boats to "stack" licenses, that is, to buy licenses for any and all of the three zones. Rather than encouraging a larger fleet of small boats, this allowed companies with large, efficient boats to buy up licenses while forcing small, labor-intensive, and more job-producing boats out of business. The fleet has been reduced, but the capacity to overfish the stocks remains high. Only now, decisions and power are concentrated in a few corporate hands centered in Vancouver.

Mifflin's successor, David Anderson, has been faced with the stark reality of a catastrophic decline in coho stocks. In the summer of 1998, Anderson imposed a complete ban on commercial fishing of coho in areas designated by red lines on the ministry's map. In yellow-marked areas, some salmon fishing is allowed. Critics point out that while Native and commercial fishers have been hard hit by the coho policy, Washington State fishers have been allowed to continue to take coho in the Strait of Juan de Fuca, where they are on the way to the Fraser River. Anderson got an American agreement to reduce coho catch by allowing them to increase their catch of chinook. DFO's egregious management strategy continues, even though there are better alternatives.

One Farmer Really Close to the Soil

OUR RELATIONSHIP WITH FOOD, LIKE OUR NEED FOR CLEAN AIR and water, should be a constant reminder that we are biological beings. But today, air is often filtered, warmed, cooled, or humidified in our homes, offices, and vehicles, and we consume far more liquid in various kinds of drinks than as just plain water. Every bit of our nutrition is plant or animal, yet people today have little appreciation of the biological nature of their food.

The meals we consume today seem disconnected from the Earth, where they originate. The farmer and writer Wendell Berry once told me that in North America, on average, food is consumed 3,200 kilometers (1,988 miles) from where it is produced. But because air, water, soil, and biodiversity are economic "externalities," the ecological consequences of global trade are not reflected in the cost of food. When I inquired why New Zealand–grown rather than Ontario lamb was featured on the menu of a fancy restaurant north of Toronto, the answer was "It's cheaper." The true ecological cost of fresh fruit and vegetables in winter in a northern country like Canada is never revealed in their price.

For most of history, food was ingested close to where it grew. We ate locally and seasonally. Our severance from an immediate and intimate relationship with nutrition is a direct result of the disconnection from land that characterizes modern urban society. Food is one of the best ways to reassess the way we live.

In the farm community of Sakurai City in Nara prefecture, I encountered fifty-two-year-old Yoshikazu Kawaguchi, who practices a radically different kind of agriculture called natural farming. Kawaguchi begins with the understanding that nature is a complex community of living things that humans do not understand. Consequently, one or a few species of plants or insects can't be defined as "good" or "bad" when we know so little about their roles in the entire ecosystem.

Kawaguchi doesn't use chemical fertilizers or pesticides, and what's more, unlike organic farmers, he does not till the soil. He lets the "weeds" that are competing with his crops grow, or when he does intrude, he cuts the plants above ground and leaves their tops on the earth.

Within the plant cover, onion shoots were poking through, and potatoes inserted into slits were sprouting roots. Wheat seeds are cast over the paddies and harvested in early June. Then rice is planted and harvested in November. Kawaguchi's soil is sticky, black, and pungent with decaying vegetation, in contrast to his neighbor's neat, plowed, and weedless furrows of gray dirt.

For twenty years, Kawaguchi was a typical farmer using chemical pesticides and fertilizers. Then, about twenty years ago, his family began to get sick repeatedly and he developed a life-threatening liver disorder. He happened to read a series of newspaper articles on complex pollution, which made him realize that his family's health problems might be caused by the chemicals he was using. He then read *The One Straw Revolution,* a seminal work by Masanobu Fukuoka on natural farming, which prescribes sowing seeds onto unplowed ground. By following Fukuoka's methods, Kawaguchi lost his entire rice crop in the first two seasons. But with tenacity and observation, he succeeded the third year. "I realized that natural farming methods are not fixed. The natural farmer should be constantly flexible and must learn intimately about soil, insects, and natural conditions of the area."

Once he had changed methods, Kawaguchi looked at the world through different eyes: "It was only after I started natural farming that I felt happy to be a farmer. Before, I felt as if I was standing in a deadly world. My rice and vegetables were growing, but I watched insects dying in agony and my fields became silent places devoid of any other forms of life."

Kawaguchi believes the increased wealth created by farming with machines and chemicals is an illusion. After paying for machinery, fuels, and the ever-increasing amount and variety of fertilizers and pesticides, much of the benefit of large-scale agriculture vanishes. Furthermore, the yields may be high, but Kawaguchi calls the food tasteless and devoid of adequate nutrition. And his most pointed criticism is that modern farming breaks apart the web of life and threatens people's health.

For Kawaguchi, his farming methods have become a spiritual way of life. He has tried meditation and other forms of spiritual discipline, but he ultimately realized that farming is his vocation and that nature is his teacher. He told me, "We have been seeing other life-forms as our enemies. But if we see them as friends, it changes how we act. The more we learn about what's happening in soil, the more we learn about life."

That's what is distinctive about natural farming. Instead of trying to impose a human agenda on nature, natural farmers know there is much that they still have to learn, so they try to let nature guide them. He says, "The land lets you live, the seasons give you the food from the land." Kawaguchi recognizes that humans are no longer hunter-gatherers living by the dictates of nature. The very acts of collecting seeds and planting them, whether by mimicking nature and casting them on the ground or by using large machines and chemicals, are deliberate. "I cut plants competing too much with my crops," he says without apology. "I select seeds too. I am a farmer, not a gatherer." But he begins from an understanding that categories such as "weed," "pest," "good," or "bad" are human definitions, not meaningful biological classifications.

We don't know all of the constituents of soil, air, or water; nor do we understand how they interact or maintain the Earth's productivity and resilience. Thus, one must begin with respect for the 3.5 billion years during which life evolved without human intervention. Protection of the integrity of the soil ecosystem and the air and water that nourish it is uppermost in the priorities of a natural farmer, because as long as they are maintained, so is human life. "We have to go along with nature, never try to impose our formula," says Kawaguchi.

But farming in Japan, as in North America, has changed. "Enjoying nature is not part of mainstream farming. It has become an enterprise designed only to make a lot of money. Farmers are removed from the philosophy of raising life. It's become totally scientific. So they are surrounded by nature, but they are also removed from it at the same time."

For Kawaguchi, science has been a major cause of this change:

> Scientific western thinking puts man and nature in conflict. It says nature can't do it properly without man. In natural farming, humans are seen to be a part of nature. The yield is smaller with natural farming, but the food is real; it has more life. It's not artificially pumped up. You need less of it to live. The ideal situation is that you grow what you eat, [and] that you eat what is grown in your area.
>
> Human knowledge is so limited. So real human knowledge must achieve a kind of *sattori* [enlightenment]. What science can find is part of something, but just a part.... Each organism has a wide range of activities, so you can't just pull out one function; there are many more organisms than science knows, and each one is complete in its existence and part of a totality. So to find just one wonderful function of one organism and bring it into the field is just disrupting the harmony of the field.

Kawaguchi pointed out that most approaches to farming focus on one or two elements of numerous possibilities. Thus organic farming emphasizes natural fertilizers and no chemical pesticides, permaculture focuses on cultivating native species, and an approach called "effective microorganisms" uses soil microbes to counter oxidation, which breaks things down. But ultimately we have to pay attention to the whole complex of soil, air, and water and the balance of life within it. As Kawaguchi stresses, "The basic thing is to trust life and let it live in the natural world."

He also predicts:

> The scientific and technological society has been developed, but life is worsening. Yet the mainstream is still blind and rushing along. It's certain this civilization has to collapse. Within this troubled world, a new civilization is already beginning. The new civi-

lization that is sprouting is based on the value of life and chooses to live in harmony with the Earth.

We have to get rid of our obsession with death. We must let nature do its work and trust the body. When a crisis happens, we have to respond. We take a narrow view, but we have to be calm. We must accept our mortality, but do not give in to death by disease. By accepting death, this is the only way to accept life. If we try to escape death we are actually denying life. That state works against all life. By accepting death, then you can live. So when we are ill with disease, we must accept that but not give in.

We would do well to study the philosophy of natural farming. Most of us live in cities away from the primary production of food. But Kawaguchi's reflections about farming inform us of a different way of seeing our place in nature, a way that might guide us into a balance with the things that make all life possible.

Epilogue

EVER SINCE RACHEL CARSON CATAPULTED THE ENVIRONMENT INTO public consciousness, our awareness of ecological degradation has risen steadily. Fueled by accidents and discoveries like CFC destruction of ozone, Three Mile Island and Chernobyl, the *Exxon Valdez* oil spill, chemical fires and spills at Seveso, Bhopal, and Basel, the largest gathering of heads of state in history took place in Rio de Janeiro in June 1992 at the Earth Summit. It was supposed to signal a turning point in history: henceforth, the environment could never be ignored in political and economic decisions and programs. Sustainable development was the rallying theme and Agenda 21 the blueprint of its implementation.

But the 1990s were also a period of explosive economic growth fueled by the dotcom bubble, the information superhighway, globalization, and biotech industries. In their preoccupation with profits, GDP, and economic expansion, business and political leaders alike came to view environmental assessments and regulations as a nuisance and a barrier to continued growth. Despite all the publicity at Rio, Agenda 21 was very quickly declared too expensive and ignored by the industrialized nations.

In my personal encounters with people across North America and Australia, however, I sensed a public concerned about the health threats to their children from air and water pollution and contaminated food. Over and over, I was asked, "What can I do to make a real difference?" It is my feeling that people are ready to take responsibility but don't want to feel foolish by trying while others are indifferent, or by taking steps that are simply token gestures but don't add to any significant change.

So the David Suzuki Foundation formed a partnership with the Union of Concerned Scientists (UCS) to seek the areas where individual citizens do significantly affect the environment. The two groups found that our lifestyles affect climate, through the greenhouse gases released on our behalf; air and water, through pollution; and habitat degradation, through urban sprawl. They also found that the most important activities influencing those four areas (greenhouse gases, air pollution, water pollution, and habitat degradation) are where we live, what we eat, and how we move about. We need housing, food, and transportation, and it is here that we can lighten our ecological footprint. The UCS and DSF developed a set of ten steps that can be taken by ordinary citizens and, if taken by tens of thousands of people, will add up to a significant reduction of impact.

We have called the ten steps "The Nature Challenge," and we are asking people to make a commitment to carry out at least three of them in the coming year. Not only could such a commitment have a perceptible impact, but if enough people sign on to such a commitment, it could be the movement that would compel the political and business communities to join. That is how social change comes about.

You can find background information about the Nature Challenge and sign up at www.davidsuzuki.org. Here are the suggested ways you can lighten your impact on the planet and nature:

The Nature Challenge

- Find ways to reduce your home heat and electricity use by 10 percent.
- Replace household pesticides with nontoxic alternatives.
- Choose an energy-efficient home and appliances.
- Eat meat-free meals once a week.
- Prepare your meals with locally grown food for a total of one month a year.
- If you buy a car, make sure it's low polluting and fuel efficient.
- Use transit, carpool, walk, or bike one day a week.
- If you move, choose a home within walking or biking distance from your daily destinations.
- Support alternatives to the car by urging improved public transit and walking and bike paths.
- Learn more about conserving nature and tell your family and friends and political and business leaders.

References

INTERCONNECTIONS

There's a Lot to Learn

Winchester, N.N. 2002. Temperate rainforest canopies: Gardens in the sky. *Wildflower* 18:34–37.

Erwin, T. 1982. Tropical forests: Their richness in *Coleoptera* and other arthropod species. *Coleopterists' Bulletin* 36:74–75.

—Update

Venter, J.C., *et al.* 2001. The sequence of the human genome. *Science* 291:1304–51.

Li, W. H., et al. 2001. Evolutionary analysis of the human genome. *Nature* 409:834.

All Species Foundation. <http: / / www.all-species.org.>

The Thrill of Seeing Ants for What They Are

Anderson, J.B., M.L. Smith, and J. Bruhn. 1992. The fungus *Armillaria bulbosa* is among the largest and oldest living organisms. *Nature* 356:428–31.

Grant, M. 1993. The trembling giant. *Discover* 10:82–89.

—Update

Ferguson, B.A., T.A. Dreisbach, C.G. Parks, G.M. Filip, and C.L. Schmidt. 2003. Coarse-scale population structure of pathogenic *Armillaria* species in a mixed-conifer forest in the Blue Mountains of northeast Oregon. *Canadian Journal of Forest Research* 33:612 –33.

The Case for Keeping Wild Tigers

Mowat, F. 1985. *Sea of slaughter.* Toronto: Seal.

How Little We Know

Wilson, E. O. 1987. The arboreal ant fauna of Peruvian Amazon forests: A first assessment. *Biotropica* 19:245–51.

Food and Agriculture Organization of the United Nations (FAO). 1990. Forest resources assessment, 1990: Tropical countries. FAO Forestry Paper 112. Rome: FAO.

—Update

Wilson, E.O. 1992. *The diversity of life.* New York: W.W. Norton & Co.

A Walk in a Rain Forest

Hallé, F. 1990. A raft atop a rainforest. *National Geographic,* October.

Global Warming

Marshall Institute report referred to in editorial of *Globe and Mail,* 12 October, 1990.

Intergovernmental Panel on Climate Change. 1995. *The science of climate change.* Edinburgh: Cambridge University Press.

Why We Must Act on Global Warming—Update

Malcolm, J.R., C. Liu, L.B. Miller, T. Allnutt, and L. Hansen. 2002. *Habitats at risk: Global warming and species loss in globally significant terrestrial ecosystems.* Report for the David Suzuki Foundation and World Wildlife Fund.

Kling, G.W,. *et al.* 2003. *Confronting climate change in the Great Lakes region: Impacts on our communities and ecosystems.* Vancouver: Union of Concerned Scientists and the David Suzuki Foundation.

Torrie, R. 2001. *Power shift: Cool solutions to global warming.* Vancouver: David Suzuki Foundation.

Torrie, R., R. Parfelt, and P. Steenhof. 2002. *Kyoto and beyond: The low emission path to innovation and efficiency.* Vancouver: David Suzuki Foundation and the Climate Change Action Network.

Ecological Footprints

Wackernagel, M., and W. Rees. 1995. *Our ecological footprint: Reducing human impact on the Earth.* Gabriola Island: New Society Publishers.

—Update

Wackernagel, M., N.B. Schulz, D. Deumling, A.C. Linares, M. Jenkins, V. Kapos, C. Monfreda, J. Loh, N. Myers, R. Norgaard, and J. Randers. 2002. Tracking the ecological overshoot of the human economy. In *Proceedings of the National Academy of Sciences of the United States of America* 99:9266–71.

A Boost for Biodiversity

Tilman, D., and J.A. Downing. 1994. Biodiversity and stability in grasslands. *Nature* 367:363–65.

ECONOMICS AND POLITICS

The Hubris of Global Economics

Faraclas, N. 1997. Critical literacy and control in the new world order. In *Constructing critical literacies: Teaching and learning textual practice,* edited by S. Muspratt, A. Luke, and P. Freebody. Creeskill, NJ: Hampton Press.

Ecologists and Economists Unite!

Simon, J. 1996. *The ultimate resource 2.* Princeton: Princeton University Press.

A Progress Indicator That's Real

Kennedy, Robert, Jr. 1994. Quoted in C. Cobb, and T. Halsted, *The genuine progress indicator: Summary of data and methodology* (San Francisco: Redefining Progress Institute).

Conable, Barber. Quoted in Cobb and Halsted, ibid.

Endless Growth—An Impossible Dream

Brundtland, G.H., ed. 1987. *Our common future.* World Commission on Environment and Development. Oxford: Oxford University Press.

Three Economists

Simon, J. 1981. *The ultimate resource.* Princeton: Princeton University Press.

Daly, H. 1991. *Steady state economics.* Washington, DC: Island Press.

Daly, H., J.B. Cobb, and C.W. Cobb. 1989. *For the common good: Redirecting the economy toward community, the environment and a sustainable future.* Boston: Beacon Press.

Economics and the Third World

George, S. 1976. *How the other half dies: The real reasons for world hunger.* Harmondsworth, Middlesex: Penguin Books.

George, S. 1988. *A fate worse than debt: A radical analysis of the Third World debt crisis.* Harmondsworth, Middlesex: Penguin Books.

Consumption as a Deliberate Goal

Wachtel, P.L. 1988. *The poverty of affluence: A psychological portrait of the American way of life.* Gabriola Island: New Society Publishers.

Kanner, A.D., and M.E. Gomes. 1995. The all-consuming self. *Adbusters* (summer).

Lebow, V. 1960. Quoted in Vance Packard, *Wastemakers* (New York: Van Rees Press).

Assigning a Value to Nature

Schumacher, E.F. 1974. *Small is beautiful: A study of economics as if people mattered.* London: Abacus.

Constanza, R., R. d'Arge, R. de Groot, S. Farber, M. Grasso, B. Hannon, K. Limberg, S. Naheen, R.V. O'Neill, J. Paruelo, R.G. Rasbin, P. Sutton, and M. van den Belt. 1987. The value of the world's ecosystem services and natural capital. *Nature* 387:253–60.

Toward More National Economies

Daly, H. 1994. Farewell lecture to the World Bank, 14 January.

Following a Different Path

Mutang Urud, A. 1992. Address to the United Nations General Assembly, New York City, 10 December.

The True Price of a Tree

Suzuki, D.T., and H. Dressel. 2003. *Good news for a change: Hope for a troubled planet.* Vancouver: Greystone.

Plundering the Seas—Update

Earthtrust and driftnets: A capsule history from 1976–1995. <http://www.earthtrust.org/dnwcap.html>.

Stump, K., and Batker, D. 1996. *Sinking fast: How factory trawlers are destroying U.S. fisheries and marine ecosystems.* Greenpeace USA report. <http://archive.greenpeace.org/~usa/reports/biodiversity/sinking_fast/>.

Linden, E. 1999. An ill tide up north. *Time*, 16 August, 53–54.

Watling, L., and E.A. Norse. 1998. Disturbance of the seabed by mobile fishing gear: A comparison to forest clearcutting. *Conservation Biology* 12:1180–97.

SCIENCE, TECHNOLOGY, AND INFORMATION

Biotechnology: A Geneticist's Personal Perspective

Roszak, T. 1974. The monster and the titan: Science, knowledge and gnosis. *Daedalus* 103:17–32.

East, E.M., and D.F. Jones. 1919. *Inbreeding and outbreeding: Their genetic and sociological significance.* Philadelphia: Lippincott.

Jensen, A. 1969. How much can we boost IQ and scholastic achievement? *Harvard Educational Review* 39:121–35.

Bodmer, W., and L. Cavalli-Sforza. 1970. Intelligence and race. *Scientific American* 223:19–29.

Suzuki, D.T. 1986. *Metamorphosis: Stages in a life.* Toronto: Stoddart.

Suzuki, D.T. 1989. *Inventing the future.* Toronto: Stoddart.

Suzuki, D.T. 1994. *Time to change: Essays.* Toronto: Stoddart.

Suzuki, D.T., and P. Knudtson. 1990. *Genethics: The ethics of engineering life.* Cambridge: Harvard University Press.

Perlmutter, R. 2003. *Maclean's*, 16 June, 62.

Affleck, L. 2001. Crops and other field uses. In *New Zealand Royal Commission on Genetic Modification.*

Strohman, R. 2000. Crisis position. *Safe Food News.*

Venter, C. 2000. Quoted in The genome warrior, *New Yorker*, 12 June.

Venter, C. 2001. Quoted in Why you can't judge a man by his genes, *Times*, 12 February.

Ewen, S.W.B., and A. Pusztai. 1999. Effects of diets containing genetically modified potatoes expressing *Galanthus nivalis* lectin on rat small intestine. *The Lancet* 354:1725–29.

References

A Humbling Message of Ants and Men

Wilson, E.O. 1996. *Naturalist*. Washington, DC: Island Press.

Wilson, E.O. 1984. *Biophilia: The human bond with other species*. Cambridge: Harvard University Press.

Kellert, S.R., and E.O. Wilson, eds. 1993. *The biophilia hypothesis*. Washington, DC: Island Press.

The Really Real

Ornstein, R., and P. Ehrlich. 1999. *New world, new mind: Moving toward conscious evolution*. Carmichael, CA: Touchstone Books.

The Hidden Messages

Roszak, T. 1986. *The cult of information: The folklore of computers and the true art of thinking*. New York: Pantheon.

Kanner, A.D., and M.E. Gomes. 1995. The all-consuming self. *Adbusters* (summer).

Stöll, C. 1995. *Silicon snake oil: Second thoughts on the information highway*. New York: Doubleday.

Novak, P. 1988. Commencement address at Dominican College, San Rafael, CA. Published in *Dominican Quarterly* (summer).

Wolf, G. 1996. Steve Jobs: The next insanely great thing. *Wired*, February. <http://www.wired.com/wired/archive/4.02/jobs.html>.

Kay, A. 1995. U.S. Congress joint hearing on educational technology in the 21st century, Science Committee and the Economic Opportunities Committee, U.S. House of Representatives, 12 October.

Postman, N. 1995. Quote from a speech given in town hall in New York City. *Utne Reader*, July/August.

Clinton, U.S. President William (Bill). 1995. Quoted in *New York Times*, 23 September.

Shenk, D. 1994. Investing in our youth. *Spy*, July/August.

Shenk, D. 1997. *Data smog: Surviving the information glut*. San Francisco: HarperEdge.

Are These Two Reporters on the Same Planet?

Desbarais, P. 1994. *Maclean's*, 25 April.

Kaplan, R.D. 1994. The coming anarchy. *Atlantic Monthly*, February.

Gee, M. 1994. Apocalypse deferred. *Globe and Mail*, 9 April.

Why a Warmer World Won't Be a Better World

Kendall, H., and D. Pimentel. 1994. Constraints on the expansion of global food supply. *Ambio* 23:198–205.

SCIENCE AND ETHICS

Genetics after Auschwitz

Haberer, J. 1972. Politicization in science. *Science* 178:713–24.

Müller-Hill, B. 1987. Genetics after Auschwitz. *Holocaust and Genocide Studies* 2:3–20.

Gold, R. 1989. Gene designers. *Globe and Mail*, 17 February.

The Temptation to Tamper—Update

Healthy-but-short kids may get growth hormone. 2003. *Vancouver Sun*, 8 July.

The Pain of Animals

Goodall, J. 1987. A plea for the chimpanzees. *American Scientist* 75:574–77.

Redmond, I. 1988. *BBC Wildlife Magazine*, April.

Sneddon, L., V.A. Braithwaite, and M. Gentle. 2003. Do fish perceive pain: Evolution of vertebrate *Nociception*. In *Proceedings of the Royal Society of London Series B*, 270:115.

A BIOCENTRIC VIEW

Why the Bravest Position Is Biocentrism

Devall, B., and G. Sessions. 1986. *Deep ecology: Living as if nature mattered.* Salt Lake City: Peregrine Smith.

Making Waves

Carson, R. 1962. *Silent spring.* Boston: Houghton Mifflin.

The Invisible Civilization

Ellison, R. 1952. *Invisible man.* New York: Random House.

LEADERS, ROLE MODELS, AND SUCCESS STORIES

A Heroic Shepherd

Davis, W., and T. Henley. 1990. *Penan: Voices for the Borneo rainforest.* Vancouver: Western Canada Wilderness Committee.

Child Power—Update

Cullis-Suzuki, S. 1993. *Tell the world: A young environmentalist speaks out.* Toronto: Doubleday.

One Logger and His Forest

Maser, C. 1988. *The redesigned forest.* Toronto: Stoddart.

Loomis, R. 1988. *Wildwood: A forest for the future.* Gabriola Island: Reflections.

A New Kind of Political Leader

Gore, A. 1992. *Earth in the balance: Ecology and the human spirit.* New York: Houghton Mifflin.

Fisheries That Flourish

Walters, C. 1995. *Fish on the line: The future of Pacific fisheries.* Vancouver: David Suzuki Foundation.

Pinkerton, E., and M. Weinstein. 1995. *Fisheries that work: Sustainability through community-based management.* Vancouver: David Suzuki Foundation.

One Farmer Really Close to the Soil

Suzuki, D.T., and K. Oiwa. 1996. *The Japan we never knew: A journey of discovery.* Toronto: Stoddart.

Index

Credits

"Catching an Epiphany" was previously published in *When the Wild Comes Leaping Up: Personal Encounters with Nature,* edited by David Suzuki (Vancouver: Greystone Books, 2002).

The following essays were previously published in *Inventing the Future: Reflections on Science, Technology and Nature* by David Suzuki (Toronto: Stoddart, 1989): "The Power of Diversity," "Owning Up to Our Ignorance," "The Ecosystem as Capital," "Ecologists and Economists Unite!" "It Always Costs," "The Illusory Oil Change," "Nuclear Menus (Or, Eating in the Nuclear Age)," "The Prostitution of Academia," "Infoglut and Its Consequences," "Misusing the Language" (originally part of "Owning Up to Our Ignorance"), "The Really Real," "Genetics after Auschwitz," "The Final Dance on Racism's Grave," "Through Different Eyes," "The Temptation to Tamper," "The Pain of Animals," "Are There No Limits?" "Borrowing from Children," "Making Waves," "Teaching the Wrong Lessons," "Losing Interest in Science," "The System and the Ecosystem," "The Invisible Civilization"

The following essays were previously published in *Time to Change* by David Suzuki (Toronto: Stoddart, 1994): "London in My Life," "Galápagos," "Human Borders and Nature," "There's a Lot to Learn" (originally titled "There's Lots to Learn"), "Elephants of the Sea," "How Little We Know," "Megadams," "Global Warming," "Near the End of Life" (originally titled "Dad's Death"), "Economic Fallacy," "Endless Growth—An Impossible Dream," "Three Economists," "Economics and the Third World," "Plundering the Seas," "Shifting Political Perspectives" (originally titled "Political Response Is Flawed"), "Science and Technology Are Still in Their Infancy" (originally titled "Science and Technology"), "Television's Real Message," "Haida Gwaii and My Home," "The New Leaders," "A Heroic Shepherd," "Grass-Roots Groups," "A Woman in Science," "Young People," "Monteverde and Children," "Child Power," "Germany—An Inspiring Example," "Water and a Canadian Scientist," "One Logger and His Forest," "Philosopher-King," "A New Kind of Political Leader"

The following essays were previously published in *Earth Time* by David Suzuki (Toronto: Stoddart, 1998): "The Thrill of Seeing Ants for What They Are," "The Case for Keeping Wild Tigers," "A Walk in the Rain Forest" (originally titled "Lessons from a Walk in a Rainforest"), "Why We Must Act on Global Warming," "Ecological Footprints" (originally titled "Stark Facts on Ecological Footprints"), "A Boost for Biodiversity" (originally titled "Study Gives Biodiversity a Boost"), "Learning from Nature" (originally titled "Lessons Taught by Nature"), "The Hubris of Global Economics," "A Progress Indicator That's Real," "Consumption as a Deliberate Goal," "Assigning a Value to Nature" (originally titled "Economics and the Real World," "Toward More National Economies" (originally titled "Economists See Errors of Our Ways"), "The *Wall Street Journal's* Insane Criteria" (originally titled "The *Wall Street Journal*'s Criteria Are Insane"), "Following a Different Path" (originally titled "The Buzzsaw of 'Progress' Hits Sarawak"), "Lessons from Humanity's Birthplace," "Live by the Box, Perish by the Box," "A Humbling Message of Ants and Men," "Virtual Reality," "The Hidden Messages," "Are These Two Reporters on the Same Planet?" "Why a Warmer World Won't Be a Better World," "Why the Bravest Position Is Biocentrism," "A Buddhist Way to Teach Kids Ecology," "Why Sterile Schoolyards Are a Waste," "Reflections While Backpacking," "Village Power Wins Victories in India," "Fisheries That Flourish" (originally titled "Study Focuses on Fisheries That Flourish"), "One Farmer Really Close to the Soil"

The following essays have not been previously published in book form: "What Can I Do?" "Arrival of an Alien," "Learning to Slow Down," "The True Price of a Tree," "True Wealth," "Biotechnology: A Geneticist's Personal Perspective"

The following sources have given permission for quoted material:

From "How Big Is Our Ecological Footprint?" by Mathis Wackernagel, 1994. Pamphlet may be obtained from Department of Family Practice, University of British Columbia, Vancouver, B.C.

From "Critical Literacy and Control in the New World Order" by Nicholas Faraclas, in *Constructing Critical Literacies,* edited by Sandy Muspratt, Alan Luke, and Peter Freebody, Hampton Press, 1997.

From *Adbusters*, volume 1, number 3.

From *The Ultimate Resource* by Julian Simon, 1992.

From "Farewell Lecture to the World Bank" by Herman Daly, January 14, 1994.

From an address to the United Nations General Assembly to launch the UN Year of Indigenous People by Anderson Mutang Urud, December 10, 1992.

From *Naturalist* by Edward O. Wilson, 1996, Island Press.

From *Biophilia: The Human Bond with Other Species* by Edward O. Wilson, 1984, Harvard University Press.

From Thomas Veltre at the Wildlife Filmmakers' Symposium in Bath, England, 1989.

From "Genetics after Auschwitz" by Benno Muller-Hill, in *Holocaust and Genocide Studies,* volume 2, number 1, 1987.